The New Ether, Science and Speculation

Clyde H. Lane

authorHOUSE®

AuthorHouse™
1663 Liberty Drive
Bloomington, IN 47403
www.authorhouse.com
Phone: 1-800-839-8640

First published by AuthorHouse 6/19/2009

ISBN: 978-1-4389-6461-4 (s)

Printed in the United States of America
Bloomington, Indiana

This book is printed on acid-free paper.

Contents

About the Author

Clyde H. Lane has spent his life pursuing knowledge, particularly in science, the field he loves. As a child he was a fan of Dr. Huer. Quickly he noticed that Huer possessed the real power in the sci-fi fantasy, Buck Rogers. Whenever Rogers got into serious trouble, Huer bailed him out. That is power. By nine young Clyde was pondering deep questions like, if you reach the edge of the universe, what lies on the other side. His fundamentalist Baptist father, when confronted with the question responded, "Don't think about it son, it will drive you crazy". He meant well, but that was like telling a dieter to not think about food. Older folks have so many ways of discouraging the natural curiosity of their children.

Clyde grew up with grandparents following the death of his mother, when he was two years old, and the breakup of the home five years later. His first stepmother had left and his Dad's greenhouse and florist business went bankrupt soon after. Clyde's grandfather, Robert Lane, was employed by Eastman Kodak Co. Rob or Bob, as he was known, only had an eighth grade education, but he was interested in the world around him. His description of the chemistry department, where he sometimes cleaned the floors, was intriguing. Chemistry sounded like magic. The Gilbert chemistry set Clyde had was used for mischievous purposes however, stink bombs, fuses, and explosives – what else? In Cub Scouts the boys constructed and flew kites and built their own crystal radios from scratch. Clyde loved the hands on activity. He was interested in everything, but he did not focus well. Day dreams about great adventures occupied his mind. The young mind is surprisingly prescient. Laney predicted to his buddies, that the United States would land a man on the moon by rocket ship before 1970, when they would be forty years old. Of course they thought him a fool.

One of his favorite books was entitled, "The Book of Marvels". Each entry was a source of excited wonder for this curious lad. He loved "Believe It Or Not" by Ripley in the local paper and read Flash Gordon and Capt. Marvel comics. School science was okay, but it didn't grab him like the popular science book, "Human Destiny" or later "The Birth and Death of the Sun" or "One Two, Three, Infinity". George Gamow brought more people into science than hundreds of science teachers. Many young folks also got interested because of science fiction classics. School science was not made particularly interesting, unfortunately. It should be and it could be.

After high school, Mr. Lane worked various jobs, including six months in Bausch and Lomb's glass research laboratory. This fascinating job encouraged the confused, still searching young fellow to consider a career in materials science. But affairs of the heart and wanderlust took him first to Miami Beach then to Butte, Montana. There Uncle Sam caught up with him and he spent the next two years as a paratrooper with the 11th Airborne Division. Following that adventurous period, Clyde married, settled down, and soon enrolled in night school at Utica College of Syracuse University on the G.I. Bill, undoubtedly one of the most influential pieces of legislation ever to have emanated from the halls of Congress.

College was Clyde's cup of tea. Old enough now to appreciate the value of formal education, he took to it like a duck to water. After completing the chemistry curriculum, save Physical Chemistry, which the college would not teach at night for lack of students, he switched his major to Physics to complete his degree. As often happens, although a blow at the time requiring an extra year of night school, the switch was a blessing in disguise. Mr. Lane had been working as an assistant chemist for Rome Manufacturing Co. the maker of copper plated Revere Ware. Following graduation in June 1960, a physicist job became available at Rome Air Development Center, a research facility of the U.S. Air Force. The position was in a group concerned with the reliability of electronic devices. It became the Microelectronics Laboratory at R.A.D.C. where both hybrid and silicon integrated circuits were fabricated and failure mechanisms studied. During Mr. Lane's twenty-eight year distinguished career at the center, he earned Master's degrees in Solid State Science (Syracuse University, 1970) and Systems Management (University of Southern California, 1974) and taught Advanced General Physics for two semesters at Utica College.

Mr. Lane has published and presented many research papers in recognized, refereed journals, such as IEEE Trans on Electron Devices, Metallurgical Transactions of the American Society for Metals, and the IEEE Trans on Reliability Physics. He authored the first paper connecting interface stress to interface electronic states in metal-oxide-semiconductor devices, having discovered the

4

interfacial stress while engaged in fabricating a dielectrically insulated integrated circuit. He went on to connect the stress and associated interface states to ionizing radiation effects and managed the generation of total ionizing radiation dose and neutron radiation specifications for MOS devices used in space applications. His in-house research made particular contributions in the understanding of electro-migration. Broad knowledge of failure mechanisms and accelerated testing led to articles on nichrome corrosion, time dependent dielectric failure, visual inspection reliability, electrostatic discharge and reliability modeling. He holds several patents for electron devices. In addition to his research pursuits, Mr. Lane represented R.A.D.C. on the national Advisory Group on Electron Devices, held various posts for the Reliability Physics and Electrostatic Discharge Symposia as well as GOMAC, a yearly government sponsored research conference.

Since his retirement, he has continued to follow developments in science. His major disappointment has been the inability to get his most important discovery published. In the late 70's he realized that quantum mechanics is a natural development from the kinetic theory of gases, if one allows that the universe is filled with sub-electron particles. He also had the temerity to point out that the twin paradox has never been resolved and that the Michelson-Morely experiment did not disprove the ether, as commonly alleged. Despite notable contributions in his professional career, the mere suggestion that vacuum was not empty brought the crank label.

For a scientific community that prides itself on rational inquiry and the rejection of the cult of personality, this knee jerk response which refuses to address legitimate questions is intolerable. Until the technical questions raised by his work are answered, Mr. Lane will continue to put forward his ideas in the hope that someone will, one day, actually address the physical model being proposed and demonstrate why it is invalid.

That is why the little book has been written. In almost thirty years, he has yet to encounter such an individual. The usual response addresses the quality of the writing and then dismisses the content with some remark like, that paradox has already been resolved, or Michelson-Morely proved there is no material filling space. Both responses are incorrect. One response long ago was that if particles fill space they must be photons which don't possess rest mass (dummy implied). That person would have given the same comment about neutrinos – they don't possess mass. Now we know that conclusion was wrong.

When there is an obvious problem with the paradigms of modern physics (they are not on the same page, and playing different tunes) it is surprising how readily new ideas are dismissed and how cavalierly one is treated when opposing conventional wisdom.

The author has encountered cranks, as most of us in science have. There is a good way to discriminate valid ideas or concerns from crank science; the true crank has no sense of humor, no perspective. In this extended article, the author has offered a test. Read the author's objection to the accepted resolution of the twin paradox in the section entitled, "Lorentz Equations and the Twin Paradox", see page 35. If you agree that the paradox has not been resolved, please read the rest of the article. If you still think it has been resolved, thank you for your time and good luck. For those who think I may have no sense of humor, I offer this simple poetic effort.

Fundamental Xions

Twinkle, twinkle little xion,
Jumping out of space,
Now you're here, now you're there
Why don't you stay in place?

You're always waving
And flying about,
Forever mysterious,
Your nature in doubt.

Securely entangled,
I have you at last,
Then you annihilated,
And slipped through my grasp.

So simple, so small,
 So fundamental and free,
Wonderfully hidden
In the universal sea.

Electrons and protons
Emerged from your stew,
Magically mixed in
A pervasive cool brew.

Hide no more quintessent elf,
We know now what you are
Dismissed so long ago,
Our theories strayed too far.

Introduction

Quintessence, the fifth essence, is a code word for ether. Einstein had hardly dismissed the old ether, replacing it with "empty space" when it was back again because of developments in Quantum Mechanics. Vacuum it seems was a sea of particles popping in and out of existence. And vacuum had properties like impedance that had to be matched for the most efficient transfer of electromagnetic energy. When Einstein published his equations of General Relativity, they looked just like equations governing the behavior of some dynamic elastic material.

This material, which deforms around a body possessing mass, he insisted, was "empty space", that is, nothing. But "nothing" doesn't bend, nor does it have a shape; only "something" has that property. And so, for almost one hundred years, we have lived with the ridiculous situation in which the vast majority of physicists believe the vacuum is something, often described as a fabric, as in the fabric of space-time, something that has texture and physical properties, yet none dare call it ether. That word is taboo. Anyone having the temerity to use it jeopardizes his career. Often, he is dismissed with a sneer and labeled a crank.

Many good investigators, seeing first hand the effects of vacuum and reporting them as the result of a material filling all space have been denied publication either because they used the forbidden word or were not clever enough to disguise their conclusions. This booklet is dedicated to those unknown souls who never could get published or published their own work and were thus considered misguided zealots who just didn't understand the new physics. It is disheartening to have your work dismissed without ever having a chance to discuss it because, "everyone knows' it must be incorrect. Some of the science establishment act as if they were cardinals of the church and you questioned the virgin birth or transubstantiation.

Fortunately, cracks are beginning to form in this wall of silence. Robert Laughlin, in his book, "A Different Universe" [1], lets one know that he doesn't think ether is a dirty word and forbidden concept. He writes, "ether nicely captures the way most physicists think about the vacuum". The gentleman who coined the word, quintessence for dark energy, was very clever and certainly

8

understood the connection to the old ether. Actually, I prefer the English spelling, Aether, because I associate ether with my tonsil operation and the unpleasant after effects.

The old ether was often considered a continuum, but some Greeks and Brahmins thought it composed of the finest atoms. The proposed new ether then, is the oldest ether in concept. We do not know what those finest particles are however. Because we need a name for them and x is traditionally the symbol for the unknown, I wanted to call them xons. Unfortunately, this suggests an association with the oil company, especially when one talks about xon gas. Yons sounds weird and Z is taken. Perhaps we should call them coupons, because later they can be traded in for something more appropriate. Finally, I decided to use the Greek xi for x and call them xions. The word can be pronounced (ex-e-on) if xion is too political. Excitons might not be bad, but for now let's stick with xions and get on with the story.

This little booklet is the result of a simple discovery the author made several years ago, that seemed to offer an obvious solution to the major problem of modern physics. The two pillars of modern physics, General Relativity and Quantum Mechanics are incompatible. For eighty years we have had two successful theories that do not see the world in the same way. Relativity, the more classical in a sense, is a study of the space-time continuum. Quantum Mechanics treats the universe as made up of discrete entities, driven by chance, and dependent on observers. Although both theories dismissed the absolute reference frames of classical physics and have been marginally integrated, they remain fundamentally at odds. No successful quantum gravitation theory exists, although it has been pursued by the best minds for many decades.

Quantum mechanics was a major departure from the past, although we will see later, the departure was not as radical as some have thought. A new biography of Max Born is entitled, "The End of the Certain World" [2]. That end began with Heisenberg, but Born was intimately involved, as the book points out. Such an attitude, the end of certainty, seems a little strange however, when you realize that in fact, chance, probability and uncertainty had been introduced into physics years before with the Kinetic Theory of Gases and Statistical Mechanics. A different perspective will be introduced, relative to the development of physical theories and what they mean, in this work.

Many problems in understanding the sometimes seemingly strange, physical world, offered by present day science, have to do with interpretation. Physics has never been able to divorce itself entirely from philosophy or religion. We have articles of faith like Occam's razor; the simpler of two theories, that solve a problem equally well, is assumed to be the more correct. Physical theories are interpreted based upon a philosophical point of view. As the years have gone by, the line between physics and philosophy has often become blurred as theories have become more abstruse. Imagine, professional physicists have told us common sense can no longer be trusted.

This author is opposed to the concept that only the elite can understand their world. Admittedly, one must spend some time to get familiar with a new idea and one must possess a degree of mental competence, but basically physical principles are rather simple. It is the mathematics that must be mastered to solve real problems that is difficult. What could be easier to understand than the acceleration of an object is proportional to the force that set it in motion. Yet this is the foundation of classical mechanics. What is more obvious than the electrical current in a conductor is proportional to the impressed voltage across it. Even the fundamental equation of quantum mechanics, which appears on Max Born's head stone,

(1) $$qp - pq = h/2\pi i \, ,$$

is just a mathematical expression for the simple harmonic motion of a particle or a pendulum. This will be demonstrated later for the adventurous and curious who elect to continue.

Niels Bohr contributed to the idea that quantum mechanics was somehow mysterious and beyond comprehension. He seemed to divine answers and suggested approaches that came out of the blue. People were awed by the success of his intuition. He suggested that in the micro-world, nature had become inscrutable. We can no longer understand what is going on, given our limited intellect, he seemed to say. All we can do is find rules that work and use them with our mathematics to make predictions.

As most everyone knows, Albert Einstein was not happy with these developments. He never accepted a chance driven nature lacking causal relationships and dependent on observers. Yet, Quantum Mechanics rather quickly attained the status of a mature theory and the chance nature of physical reality was accepted by the majority of professional scientists. So strange was this new reality that competent scientists began to distrust their own common sense. Feynman even suggested once that if Quantum Mechanics made sense to you, you didn't understand it. Some, I believe, got to the point of thinking, if it made sense it must be wrong.

One of America's great physicists, John A. Wheeler, became so certain of the need to integrate the observer into quantum theory that he returned to the Subjective Idealism of Schopenhauer. He seems to have literally accepted Schopenhauer's statement, "The world is my idea". We have gotten into the ridiculous mind set that allows rational, well educated professional scientists, to proclaim that Schrödinger's cat is both dead and alive until the observer resolves the wave function by looking into the box, thus determining the cat's ultimate fate. Laymen shake their heads in collective disbelief. Obviously, the common man can no longer understand the world in which he lives. No wonder many have given up and returned to their childhood universe of myth, magic, and superstition.

People we look up to as our leading authorities in the hard sciences proclaim, with straight faces, that one can penetrate an energy barrier by tunneling through it. Dr. Sivana, in the Capt Marvel comic, found the secret, as you may recall. He spouted his puzzling formulas and walked through walls with no problem. Amazingly, our scientists seemed to be correct and they provided the experiments to prove it. In the peculiar world where waves and particles are different aspects of the same reality, one no longer needs sufficient energy to surmount a barrier to appear on the other side. The model seems to have become fact for most scientists. Few even question the model anymore.

Relativity too has some odd consequences which violate common sense. Electromagnetic radiation does not need a medium in which to travel. EM waves are self propagating in "empty space". What is empty space? Nothing is the answer, the lack of anything. But, this "nothing", we are advised, possesses geometry and it isn't Euclidian. Imagine, our brightest minds are telling us that nothing curves. It bends around mass and causes light waves to follow that curvature. Doesn't it bother you, the reader, to imagine self propagating electromagnetic waves forced to deviate from their straight paths by nothing? Classical physics requires an applied force to cause a moving particle to deviate from its path, but in relativity we are asked to accept that such a force is inherent in the geometry of nothing. Again experimental proof has been provided and again laymen shake their heads in bewildered wonderment. How can they ever hope to understand? The world of modern science doesn't make sense. It is magic, like religion, it involves the supernatural.

A nagging problem in Special Relativity however, casts doubt on Einstein's interpretation of his equations. He and Lorentz had produced identical equations which were assumed to describe the same physical world. Mathematically they were equivalent, but Lorentz thought they were the result of the effect of ether on a moving body and the motion was with respect to the ether. Einstein insisted that the motion was simply relative and required no ether. Einstein won the majority to his interpretation. That equation set is now known as the Lorentz transformation. The nagging problem is called the clock or twin paradox. It is a troublesome conundrum that shows that for twins, separated and traveling at a high relative speed, usually set at near the speed of light for dramatic effect, each determines that the other is aging more slowly than him. Clearly, that cannot be. The difficulty comes from the time dilatation equation that requires time slow down for a body in motion relative to one at rest. Since each twin can be considered at rest relative to the other, each twin is aging faster than the other. Differences in clock speeds become pronounced when one approaches the speed of light. So embarrassing was the failure to resolve the paradox by the '50s, that a concerted effort was launched to rectify that situation. No one succeeded, although victory was declared when some suggested that the twin who left the earth was the one who aged more slowly, since he had to accelerate to leave the earth, turn around and return. Surprisingly, this ad hoc, unproven solution has been presented as fact in teaching programs and text books. It is partially correct, but it does not solve the paradox. It is true that inertial acceleration slows a clock, but so does gravity. This observation was made by Steven Hawking in his new book, "A Briefer History of Time", but he did not go on to show why this invalidates the accepted solution to the old brainteaser.

As Einstein demonstrated, gravitational acceleration is equivalent to inertial acceleration. This fact is conveniently forgotten by those who think they have resolved the enigma. One can readily arrange, at least in a thought itinerary, to have the departing twin never exceed one g of acceleration and yet attain a speed near that of light, turn around and return to earth. This is done by starting the trip in a region of slightly less than one g and manipulating the propulsion system and spacecraft orientation during the trip. Although never exceeding one g of acceleration, in about fifty days one would be traveling near the speed of light. You could continue at that speed for a year before slowing to make your turn. Turning in the appropriate arc at the correct speed would maintain one g on your craft. You would speed up again to your maximum speed, following the turn maneuver, and maintain that speed until you are ready to decelerate, again at one g, and return home. Since each twin has experienced the same acceleration for the period of the trip, the paradox remains unresolved.

And so we see that in spite of the mountain of experimental evidence for the correctness of both relativity and quantum theories, there is something fundamentally wrong with our models and their interpretation. This booklet shows the reader what went wrong, when it happened, and why. More important, the author offers a new interpretation of existing facts and a new model which solves many long standing problems.

Entrenched ideas are very difficult to overturn, even when they don't make sense. They take on the aura of scripture. Whole lives and careers are involved, and people don't give up their fundamental beliefs easily. We like to think that people in science are persuaded by the facts. Unfortunately, that is naïve. The older generation often clings to cherished beliefs. The younger, while biased by their education, do take a fresh look. That is when change occurs. The beauty of science is that it is open to change. New investigators look at both old and new ideas with fresh eyes. There are, of course some courageous souls, who seeing a better approach, will give up a cherished old idea that can no longer be supported. Society must always be willing to tolerate the mavericks that love to challenge established ideas. They can be frustrating nuisances, but we can't afford to dismiss and even destroy those who, like Socrates, persist in asking embarrassing questions.

The model proposed in this book not only removes some old problems, but permits common sense understanding of a rational world to reassert itself. The physical world will make sense to ordinary people again. Certainly mathematics becomes no less difficult. In spite of the simplicity of Ohm's law, we need computers for complex circuit analysis. Celestial mechanics challenges our ingenuity although it is only, $F = ma$ in principle. The author's goal is to introduce more certainty through understandable analogies. That is after all the way we think. Without a good analogy most of us are confused by a new idea. Only a few unusual professionals are adept enough to read complex mathematics like a well written explanation. That does not mean uncertainty is removed entirely, but we will have a better idea of how it fits into a comprehensive view of the physical world. Without randomness we would not have certainty and order. We are sure of the refractive index of a five percent copper sulfate solution, at standard conditions of temperature and pressure. We know it will be the same every time we measure it. That certainty depends on the fact that copper and sulfate ions and water molecules that contain them are in rapid random motion. If that were not true, we would get different results in different locations and at different times. Randomness and certainty are two sides of the same coin. Therefore, one must acknowledge that certainty often depends on and emerges from chaos.

A note before proceeding: The author has elected to have the reader use Google to check some references rather than listing them by author and article in the reference section. For example, the Casimer effect is well described by a Google search, as is the Subjective Idealism of Schopenhauer and the principle used in applying Occum's Razor. Therefore, he will leave the reader to his own devices in checking some of the concepts with which the reader may not be familiar or requires a refresher or a definition. That does not mean that references are left out entirely. A traditional reference section is included for supporting articles and sources.

The Classical Physical World

Observers, ten thousand years ago, saw single objects move, sometimes by human or animal muscle, but often mysteriously. They experienced the wind, rain, flowing streams, waving leaves and witnessed celestial objects parade across the sky. The obvious conclusion - forces, seen and unseen, moved things. Forces were spirits which all living things possessed. Their source was the mysterious invisible world governed by gods. Astute observers nevertheless, began to notice relationships which allowed predictions. Astronomy was probably the first budding science which grew up with and out of religion. As Stephan Hawking says in his new book, some smart fellow must have noticed that celestial objects did their thing regardless of whether the gods were placated or not.

About five or six thousand years ago, settled communities existed where some people had enough wealth and security to spend time thinking about the nature of their physical environment. Scientist-Priests or Philosophers began to organize their thoughts and experiences into an oral tradition. By three thousand years BCE, writing was invented and that tradition was being committed to written language. Slowly the rudiments of mathematics, physics and philosophy began to emerge from religious belief systems. Somewhere in the Middle East, a people, that later split into several groups including the Greeks and Brahmins of northern India, decided that everything was made of five substances, earth, water, air, fire, and ether. We of course recognize this classification as solid, liquid, gas, plasma and vacuum. Appendix 1 contains a short paper on the ether, which has an interesting history and a peculiar impact on physics. These early scientists, natural philosophers, and priests set the tone for the development of classical physics.

Only about twenty-five hundred years ago, a few Greeks embraced a rather radical idea; perhaps the universe is rational. If true, it could be understood by logical thinkers. Some Greek natural philosophers argued that the five substances were themselves composed of more fundamental units called atoms. And so we see the beginning of the rational sciences; physics and mathematics. For mathematicians, one was the natural unit, the foundation of their investigations, for some natural philosophers it was the atom. The object, its composition and motion was the stuff of physics.

Progress was slow. There was no systematic study or State supported efforts, but gradually information was recorded and a body of knowledge accumulated. Science was the work of a few curious individuals. With the collapse of Rome and the onset of Theocracy in the West, intellectual pursuits stalled. Fortunately, classical knowledge was kept alive among the Muslim upper class of the Middle East, and the Moors and Jews of Spain. Eventually, the Greek-Roman heritage found its way back into Europe and ignited the Renaissance. Classical art, architecture, engineering, mathematics and the sciences were rediscovered or reinvented. Copernicus was not the first to conceive of the Sun centered solar system, but for the West its rediscovery was a turning point. The mere idea that the Church, and its sacred book, did not contain all truth was the crack that let in the fresh air of discovery. We all know the progress in science that followed, but the basic ideas had not changed. Physical science was still the study of states of materials and the motion of bodies.

A hint of trouble in this classical world appeared when the nature of light was investigated. Newton favored and defended the particle nature of this mysterious sub- stance. Others presented convincing evidence that light was a wave phenomenon. Which was it, wave or particle? No one at the time would have suggested both. And then there was that peculiar notion of action at a distance. Even for Newton, who suggested it, the idea was too illogical and he eventually embraced the ether, suggesting that Earth's gravity was due to a concentration gradient in the ether surrounding the earth.

Newton's universe was mechanical, a system of material objects interacting in cause-effect relationships. Physics was the study of things and their motions, an apple, a cannon ball, a bob on a string, a planet etc. A unit of mass could have any velocity. Energies were continuous and unlimited. Bodies moved smoothly, submitting to the calculus in the analysis of their motions. If you didn't have the energy to surmount a barrier, you stayed behind it. There was certainty in the universe, you could depend on logic and demonstrate it with mathematics. Why mathematics was the handmaiden of physics no one knew, but clearly it was and mathematics was the embodiment of certainty.

Such was the state of physics near the end of the nineteenth century. Fields, Kinetic Theory, Thermodynamics, and Statistical Mechanics had made their appearances, but few seemed to have thought much about the consequences of these rather new ideas.

People assumed time, whatever it was, flowed smoothly and in one direction. They lived in a world of absolutes. There was a fixed reference frame to which all motion could be referred. Some unseen Omnipotence had created this vast machine and kept it in balance. All was well with the world. In principle human beings could understand nature. Scientists of the time were the enlightened ones, often of noble birth, sometimes arrogant and condescending. But times were changing.

Little clues appeared from time to time to tell the insightful all was not well and to warn of problems to come. Many were ignored; others dismissed or set aside, like our own "twin paradox" or Schrödinger's cat. The problem of light seemed to be settled in favor of the wave model. Newton was probably wrong. Action at a distance remained controversial. Scientists of the day did battle on that concept without resolution. Unbeknownst to many, Newton himself changed sides. He could not justify the mysterious concept he had invented.

One of the first hints of trouble with the classical view, which no one seems to have picked up, was the Doppler Effect. Doppler explained the principle in 1842 and obtained test results in 1845 which verified his theory. Those involved, and Doppler himself, seem never to have appreciated the implication of his observer related frequency shifts. Relative frequency implies relative time. It should have been obvious, but as everyone knows, the obvious is often easily missed. Then, in 1887, the Michelson –Morely experiment, with its unexpected null result, was completed. The history of this experiment makes good reading. What was a United States naval officer doing in Berlin in 1881, conducting a history making experiment under impossible conditions? His initial results, although reported, were not accepted as reliable. Even the incontestable 1887 experimental results were not believed by Lord Kelvin who suggested something was amiss. The experiment was conducted again, several years later, with the same result. In the mean time, physicists FitzGerald and Lorentz both came up with ad hoc explanations for the failure of the experiment to detect the ether.

Their conclusions were that motion in the ether caused the arm of the apparatus, in the direction of motion, to contract. Despite those arguments, and much head scratching, there the problem of the ether stood until 1905.

Near the end of the century, physicists were intent on solving the blackbody radiation problem. Experimentally they had obtained a very nice temperature dependent energy versus frequency distribution for a cylindrical cavity heated to a various temperatures. Emitted radiation was channeled and measured. Investigators could accurately model the high and low portions of the curve, but they could not reproduce the experimental distribution with their equations. Motivations are lost to history, so we don't really know why Max Planck took the unusual approach of quantizing the energy in units of a constant, later given his name, multiplied by the frequency. The approach worked. His equation modeled the experimental results perfectly. Neither he nor anyone else understood what this quantum of energy was all about. Planck hated the notion of the quantum, but it worked. Perhaps this was the beginning of the "operational" approach to physics. Physicists after that began to stop seeking causes and satisfied themselves with equations that "worked", that is, they made correct predictions. If you asked why, the answer was often a meaningless wandering speculation or a curt version of, don't ask. And so we erected modern physics on quantum mechanics, general relativity, and a policy of, don't ask why, we are not capable of knowing or understanding some of the answers.

Relativity is a classical theory in the sense that it relies on a continuum. Einstein gave up on an absolute spacial reference frame and on absolute time, but that was gone with Doppler anyway, if you think about it. The General Theory of Relativity supposedly deals with the geometry of space itself, but that is a real problem. Geometry is a mathematical ideal which can be applied to a real material. Euclidian geometry is applied to local land measurements on earth. No one is fooled into thinking the earth is flat, but a sufficiently small region can be treated as if it were. We recognize the Euclidian system as an ideal model. When we say space-time is non-Euclidian, we are saying that a particular geometry, when applied to space-time allows us to make reliable calculations. The problem is, what are we applying the model to? If the answer is, "empty space", that does not answer the question, it doesn't even make sense. We all know "empty space" is a euphemism for "nothing". It makes no sense to claim that nothing has a shape. A real thing has a shape to which geometry may be applied. One must never confuse an ideal with a real. In geometry, a point can have zero width. In the real world an object always has a nonzero dimension, as does the distance between objects. Almost before Einstein got the words, empty space, out of his mouth, quantum mechanics was describing a very busy space in which "nothing." behaved very much like a sea of something.

Then there is that nasty little paradox that has plagued the Special Theory of Relativity. Contrary to popular opinion and most physicists, it was not resolved by appealing to acceleration effects. That approach does not work. The only resolution of the paradox that does work requires an absolute reference frame to which the twins' motions can be referred. At the time Lorentz developed his transformation equations, he thought the ether was the cause of the interferometer arm contraction. He thought the contraction was an actual elastic response to the force exerted by the ether. Einstein, who came up with the same equations, didn't require ether. He believed motions were entirely relative. Although the equations were the same, the physics behind them was very different. Those texts and educators who proclaim the Michelson-Morely experiment proved the ether did not exist are dead wrong. They do not know their science history. It did no such thing. If you accept Lorentz's explanation, the experiment did just the opposite; it proved ether did indeed exist. Ether was dropped because it was thought to be unnecessary. It was dropped because physicists did not clearly distinguish between the Lorentz and Einstein rationales for their equations. They thought they were equivalent or Lorentz's was a patch up approach. Therefore, by Occum's Razor, Einstein's development was viewed as most correct. As the years went by it became the only acceptable conclusion. Fortunately, the clock paradox remains a testimony to the fact that they were and are wrong. The Lorentz approach generates no paradoxes.

So entrenched is the idea that no ether exists, that even when the cosmic radiation background was detected in 1965,few reconsidered its possibility. Quantum mechanics tells us space is not empty. The Casimer effect tells us space is not empty. Vacuum particles cause a real force, measured by Steven Lamoreaux in1996. Penzias and Wilson [3] found the photon gas now designated the cosmic microwave background, CMB. Dark matter may exist, and the clincher, dark energy fills space, but none dare call it ether. Some brave soul tagged dark energy with the thinly disguised, Quintessence, the fifth state of matter, perhaps to get his paper past the censors. When Pioneer spacecrafts, 10 and 11, showed signs of being acted on by some Sun directed force, the experts scratched their heads. Lorentz would have told them what was going on - your craft are not moving through empty space. They are still flying and still slowing anomalously. We have our theories, but nature always has the last word.

Modern physics can be thought of as coming into being with the, then new, 20th Century. Planck introduced the quantum in 1900 and 1905 was Einstein's magic year. Four of his outstanding contributions that year dealt with, the Photoelectric Effect, in which he showed that photons were particles, solution of the Brownian motion problem, sometimes known as the drunk man's walk, which contributed to kinetic theory,   introduction of special relativity and his most famous equation, $E=mc^2$.

Interestingly, all four of these contributions can be tied together when we investigate the impact of an all pervasive space filling material on physics. By dumping the ether concept, Einstein gave up the chance to unify his four major contributions of that year into a single consistent theory. By insisting the vacuum was nothing, he made his search for a unified theory one that would frustrate his unique talents and render much of his later work on a unified field theory irrelevant.

The other major theory, Quantum Mechanics, developed slowly and required the insights of many contributors. Although Bohr was instrumental with his uncanny guesses, Max Born's mathematical approach laid the ground work for the new theory. His interpretation of deBroglie waves as probability waves was a key insight that seems to have been his alone. Heisenberg, of course, gave us the Uncertainty Principle. However, it was Born's understanding of matrix algebra that led Heisenberg to his Matrix Mechanics. Heisenberg was Born's pupil at Gottingen University. Dirac quickly understood Heisenberg's idea and produced his own statistical theory. This approach was not popular however, among those trained in classical physics and its associated mathematics. Introduction of probability, chance, and discontinuous functions, let alone an unfamiliar mathematics did not go down well. Bohr of course accepted it intuitively. Einstein never did accept these ideas as fundamental. The discussions between Bohr and Einstein are legion. Einstein's refusal to accept the probabilistic approach is peculiar in a way, because his solution to Brownian motion is probabilistic in nature. A closer look, in fact, shows that the elements of quantum mechanics lie in the Brownian motion problem and its solution. This statement will become clear later.

When Schrödinger presented his wave equation, many breathed a sigh of relief. Here was an approach they understood. From their training, they were comfortable with differential equations and Fourier analysis. Matrix mechanics left a bad taste even when it was shown to be identical to wave mechanics. No one likes to feel uncomfortable. It is always nice to be reminded that the brilliant are no less human than the rest of us.

Over the years quantum mechanics has given rise to quantum electrodynamics with some new interpretations and techniques. Feynman's diagrams and his "all paths" or "all histories" innovations have been major contributions to the growing understanding of the micro-world described with the aid of quantum theory. Eventually, unification of the forces would take place so that, but for gravity, we seemed to be on the way to a theory of everything.

And so the great edifice of modern physics was erected using two types of construction which were incompatible. It is a wonder the building hasn't fallen down. There are cracks and many scientists suspect the building cannot be completed this way. Roger Penrose in his book, "The Road to Reality" [4] has suggested that perhaps we need a new idea or model. Still we try, because no one has figured out what is wrong yet. This dilemma reminds one of the pyramid builders who didn't get it right the first time. The higher they went the more unstable their structure became, until eventually it partially collapsed. In defeat they learned their mistake and went on to do it right. Often we learn more by a failure than we do by success. We have been brought to a virtual standstill in our effort to quantize gravity and bring it together with the unification approach so successful with electromagnetism, the weak force and the strong force.

Unfortunately, we have lost much of the public who want to understand their world, but cannot cope with renormalization, adjusting infinities, multiple dimensions, and time travelers who can shoot their own fathers and cats that are both dead and alive at the same time. Faced with such a complex, irrational universe is it not easier to throw up your hands and say, God did it and only God understands it? Is it any wonder that many consider "intelligent design" a more logical approach to understanding the world around them, than the convoluted ramblings of those who talk in a foreign language and unwittingly invoke a condescending air? Popularizers do a commendable job, but their task is almost impossible. Without better science education and a more understandable science, we may continue to lose ground with the public.

As mentioned, there have been hints that something was amiss in our view of the physical universe. Certainly the twin paradox raised a red flag. Why physicists have been content to let it remain unresolved for almost one hundred years is difficult to understand. Somehow, over the years, we have been taught not to trust our common sense. Quantization versus continuity is another basic issue. Why is motion discontinuous? It really doesn't make sense. We are schooled to believe it doesn't have to anymore. That is a departure from the scientist's initial belief that the world was rational and that we could understand it. Now we are told to forget that, we just need equations that make the correct predictions. We don't need to understand why. Sounds like church, that's the way it is, you don't need to know why. What a great way to kill curiosity.

If anything should alert us to a fundamental flaw in our thinking it is Schrödinger's cat. The idea of a creature being both dead and alive is so bizarre as to be dismissed outright by laymen who consider such gibberish to be unworthy of their time. In philosophy, Schopenhauer's Subjective Idealism has been discredited. Why then was it accepted in physics? The only answer is that the paradox is an embarrassment which must be deflected even if the solution makes no sense. The simple fact is a cat cannot be both dead and alive regardless of the observer's input. It is as if we insisted that a tossed fair coin landed both heads up and tails up until someone looked at it. Clearly, there is something wrong with the interpretation of a theory that leads to that conclusion.

A brief digression is necessary here in order to explain how the notion of universal connectedness, provided by the xion medium solves the quantum enigma treated so well in a recent book by that name[5]. Quantum mechanics is otherworldly despite the fact that it allows extremely accurate calculations and predictions. Tunneling, wave-particle duality, and Schrödinger's cat make the point. Necessity of the observer to resolve a wave function is the unacceptable heart of the matter.

Two ideas are central to the otherworldly nature of quantum physics; one involves the reality of the microworld, the other the connectedness of the universe. If objects are real, they are assumed to be independent. They can be separated and when apart, what happens to one cannot determine what happens to the other. Action at a distance violates this idea, which is why many would not accept Newton's explanation of gravity. Ether provides the universal connectedness physics lost when it accepted the idea of "empty space" that nevertheless possessed physical properties. On pages 142 – 149 of the book, Quantum Enigma, these issues are addressed. In particular, these pages discuss Bell's theorem and inequality, published in 1965, and Stapp's later modification of that inequality. The book's authors state on page 143, "in a reasonable world, objects should be separable". I humbly disagree, to me a universally connected world is just as reasonable, perhaps more so.

Lately one has heard much about entanglement. If two entities have ever interacted with each other, they are forever after connected or entangled. Experiments have shown that twin photons, those created simultaneously with opposite polarities, can travel in opposite directions and remain entangled for as long as the experimenters could measure. Bell's inequality states that when twin photons traveling away from each other are each passed through a polarizer rotated an angle θ, but in opposite directions, the error rate in polarization determination of one photon knowing that of the other, will be equal to or less twice the error rate for rotation of a single polarizer through the same angle, symbolically, $E_R(2\theta) \leq E_R(\theta)$. If the inequality is violated, our world lacks reality or separability, or both. Quantum mechanics requires the inequality to be false. Stapp derived the same inequality without using the reality assumption. As such it addresses only separability. In actual experiments the inequality was violated. The consequence is, as a minimum, the universe must be connected.

Xion assemblies provide the connectivity, whether in fluid or solid form. Atoms are not real in the sense that they are virtual entities emergent from defects in the xion matrix. This fact does not mean observers are required however. The wave function, as a description of a particle diffusion path, is itself unreal. But there is a real path and real possible outcomes. As any person with common sense knows, the cat is either dead or alive, it cannot be both. Entangled probabilities are still just probabilities and as such, unreal compared to a physical object. A diffusing particle has a definite position after n jumps, not a probable location. The roll of dice produces a real outcome. It may not be known until observed, but the observer did not influence the outcome in any way.

Consider again that bizarre phenomenon, tunneling. If experiment and theory agree, does that mean our intuition that tunneling makes no sense is wrong, our logic misleading? No. It may mean we have an incorrect model of the phenomenon. Some experiments agreed with the predictions of the phlogiston theory of heat. That did not make the theory correct. Other experiments dictated the need for another theory or interpretation.

In spite of these problems and paradoxes, modern science has had some outstanding successes. The amazing thing about science is that it works at all. Why should we be able to understand the workings of our universe? Electrodynamics, for example, allows fantastically accurate calculations. Unfortunately, for the vast majority of the world's citizens, science is unintelligible. Logic and common sense are not enough. Today we rely on very sophisticated mathematics and high speed computers to make calculations. Because people no longer understand, they become fearful and suspicious.

The science backlash is a most regrettable phenomenon. In the United States we have the most educated population we have ever had, but knowledge of science is appalling. The majority of our policy makers don't understand much of the science and technology that is changing the world daily. Most Americans do not accept the theory of evolution. You hear, "Evolution is just a theory", but the folks that say that have no problem boarding a Boeing 757 and flying across the country, obviously unaware that the craft was designed on the computer using unproven and incomplete aerodynamic theory. A nation where the majority still believes in myth, magic, and superstition and at the same time leads the world in science and technology is schizophrenic. As scientists, we are losing support because we are not doing a good job educating the public.

To rectify the difficulties modern science has gotten into, we must return to the beginning when classical mechanics seemed to have failed and two new approaches were introduced. Those approaches were based on two facts, the need to quantize energy to generate the correct blackbody radiation distribution and the constancy of the speed of light. One replaced cause and effect with a probabilistic interpretation of mechanics and the other eliminated the absolutes classical mechanics depended on. The critical departure from the classical model came when, given a choice between Lorentz's or Einstein's interpretation of the Lorentz transformation equations, the majority sided with Einstein. Eventually, his interpretation became dogma and Lorentz's became heresy. Yes, heresy. Scientists don't burn each other at the stake, but in the words of the Japanese interrogator in a World War II film about the Dolittle raid, "We have ways of dealing with you".

What was the difference in their models? It had to do with the ether. This fifth state of matter had been with us since the ancient Greeks. The majority of interested people today believe that this material was shown not to exist by the Michelson-Morely experiment. Most physics texts make some kind of statement to that effect. They are, as previous stated, incorrect. FitzGerald and

Lorentz both realized that the null result of the experiment could be due to the ether itself. Resistance to motion in the ether could cause a body to elastically contract in the direction of motion. When Einstein published his paper detailing his new relativity concept, his equations required no ether. They were simply the result of the constancy of the speed of light. So we have two interpretations for the same set of equations. We know the consequences of dispensing with the ether, the question is, how would relativity and quantum mechanics have differed if the other option had been selected and the ether had been retained?

A reminder for those interested, a brief history of the ether is contained in a short article the author wrote for himself. It is self contained with its own references and appears in Appendix I of this work.

Analogies between Ordinary Material Behavior and Vacuum Relative to Quantum and Relativity Concepts

Momentarily we shall apply the kinetic theory of gases to the xion gas that we shall assume fills or constitutes the vacuum. This is essentially what was done in the author's 1979 unpublished article, Space-time, A Sub-electron Particle Field, but now those sub-electron particles are identified as xions, the unknown particles. Because the ideas introduced by the application of kinetic theory are so fundamental to the understanding quantum concepts, let us take a moment to discuss the history of statistical ideas and their application to the study of fluids and gases.

As discussed earlier, physics developed from an interest in how objects move. When we began to focus our attention on smaller and smaller objects, the techniques of classical physics were more difficult to use. Brownian motion, for example, can not be described following the rules of Newtonian Mechanics, it is too complicated. Chemists were among the first to require some new approach. They could measure a property of a gas called pressure, but exactly how did an assembly of particles exert a force on an object? Certainly, it was impossible to work out the mechanics of each and every particle of the gas and its interaction with other particles and with a solid surface. Enter a Belgian statistician, A. Quetelet (1796-1874)[6]. His crime and suicide data indicated constant rates over time and from country to country in Europe. He concluded that he could formulate laws of a social physics that relate to these average quantities, Maxwell may well have taken a cue from this approach with which he was familiar. The result of this statistical line of attack was the Kinetic Theory of Gases, Thermodynamics, and Statistical Mechanics, all of which were developed in the latter half of the 19th century.

Looking back, one realizes that the ideas of quantization and uncertainty are implicit in kinetic theory. In fact they are implicit in measurement itself. When one employs average values as units of measurement, he is quantizing his measurement. Take the measurement of length, we standardize on the meter as our unit, but that length can only be subdivided into some smallest practical (reliable) unit because the uncertainty in its size becomes significant. In rough carpentry, for example, we employ rulers or metal tapes marked off in sixteenths or thirty secondths of an inch. We can't accurately measure, cut, and assemble to a tighter tolerance and it is unnecessary. Our quantum of measurement, say 1/32" is not precise, it is really an average value with some standard deviation. Thus, when we measure an eight foot 2x4, eight feet is actually 32x12x8 = 3072 average values or quanta. The error in eight feet is not a problem, errors tend to cancel, but if one attempted to maintain 1/32" tolerances, errors of various kinds would render the goal impossible to attain consistently. A sixteenth of an inch is a practical quantum unit for this purpose.

It is clear in this analysis that quantization and uncertainty arise whenever we try to measure anything. But there are artificial or arbitrary units of measure and natural or implicit units of measure. For a gas at standard conditions, the natural unit of length is the mean free path and that of time is the mean time between collisions. Their ratio gives the maximum speed of sound in the gas at those conditions. For a single crystal the natural unit of length is the lattice parameter. Quantization for measurement is imposed by necessity, because of our technological limitations usually. For waves however quantization is imposed on us by the nature of waves themselves. In the case of a gas we realize that not all frequencies are possible in a pipe full of air. There is a fundamental defined by the shape and volume of the pipe. The energy of the wave would then be equal to some action, h multiplied by 1/n times the frequency. The fundamental frequency would have energy, $E = hf$, while the first harmonic would have energy, $E_2 = hf/2$. The question is, how precise is the measurement of frequency? That and the precision with which h is known will determine how precise the energy can be measured. In the case of quantization of photon energy, there is a natural quantity of action, Planck's constant that ties a wave's frequency to its energy. The frequency is quantized like the frequencies of a guitar string, a bell, or an organ pipe, in terms of a fundamental and its harmonics, but the precision of measurement is much greater than for sonic frequencies. For photons it was found that $E = nhf$. This fact helped launch Quantum Mechanics. The reason the idea of quantization was seen as mysterious was that we had assumed any value of a quantity like energy was possible. We assume a 2x4 can be any length although our ability to measure it accurately is limited. With waves or a single crystal we have a different story. Some frequencies or lengths are not allowed. There exists for a given system, a minimum wavelength and maximum frequency to which all other frequencies and wavelengths are related by integer multipliers of the minimum wavelength or integer dividers of the maximum frequency for a wave. In the case of a crystal lattice, the minimum length is the lattice parameter. A length of crystal must be an integral number of lattice parameters. One cannot have a fractional lattice length added to the surface of the material.

To reiterate, in a gas or liquid, the mean free path is the most probable value of the distance between adjacent particles. If we measure a length in terms of mean free paths, we have quantized the measurement of length. For any gas, there exists a quantum of action given by the mass of the particle multiplied by its mean free path and its maximum velocity. That velocity is the ratio of the mfp to the mtbc, mean time between collisions at fixed conditions of temperature and pressure. Using statistics automatically introduces uncertainty. We talk about probabilities, not exact values. In a gas, we know we cannot measure both the position and momentum of a particular particle simultaneously. The errors in the measured properties of an ordinary gas however, do not normally require that the uncertainty be quantified. Later, when we began to look into the behavior of a single particle of much smaller dimensions, such as an electron or consider the energy of a wave packet, such as the photon, quantifying that uncertainty became a founding principle of quantum mechanics.

In his book, "A Different Universe", Robert Laughlin refers to emergent properties and laws. He says, quote, "The laws of nature we care about emerge through collective self- organization". What is he talking about? Difficult as it may sound, the idea is simple. In a gas, temperature and pressure are emergent properties. What is the principle of self-organization here? It is none other than an old friend, the average, mean or most probable value. For the individual gas molecule, temperature and pressure make no sense. Even for a few per cubic meter, the concepts have no value. But for a density of 5×10^{20} molecules per cubic centimeter, they are repeatable, dependable, indispensable properties of a gas. As you well know, the average, mean, and most probable value may not be exactly the same, depending on the distribution, but for our purpose we can treat them as equivalent. The point is that each of the one thousand molecules per milliliter in a sample of gas, has a different speed and energy, at any given moment. The average values, however are virtually constant over time. As a first approximation, we can assume molecular velocities are distributed normally, so they have a mean and standard deviation. The actual distribution is Maxwell – Boltzmann, in which one uses the most probable or root mean square value and the variance depends on which side of the value is being considered. For a standardized normal distribution, about 67% of the molecules will have speeds and energies within one standard deviation of the mean. Suppose now we keep the volume constant and add more molecules. For a perfect gas, if we also keep the temperature the same, the pressure will increase. What has changed, other than the density and the pressure? The answer is standard deviation or variance. As density increases for a perfect gas at fixed temperature and volume, the standard deviation decreases. The distribution is still normal, but a larger fraction of the molecules have speeds and energies closer to their mean values. If the process is continued, there will be a point where our measurement capability can no longer discriminate between the mean value and one standard deviation from the mean, then two, three, etc. By the time the difference between our measurement error and the variance from the mean are equal, we will see all molecules as having the mean values for the properties we have been monitoring. Since the distribution still exists, there may be one in a million molecules with energies much higher than the mean, but they will not be seen unless special equipment is set up to detect them. Distribution tightening offers the explanation for tunneling. Later on we will see that fermions can be viewed as emergent particles from a dense xion material as can their rest masses. Distribution tightening also may account for the fact that electrons are apparently identical. The results are dramatic, yet anyone familiar with the idea of an average and a distribution can understand them.

Before we move on, I would also like to expand on the similarities between Einstein's "empty space" and ordinary solid or liquid material. In Laughlin's book, ref.1, he makes the case that many have noted the uncanny similarities between the superconducting and superfluid states and vacuum. Again, on page 121, he says, "ether nicely captures the way most physicists actually think about the vacuum". Further on he states, "Relativity actually says nothing about the existence or non-existence of matter pervading the universe, only that any such matter must have a relativistic symmetry", a point this author and many other "etherists have tried unsuccessfully to make for many many years.

There are other intriguing similarities between special relativity and the behavior of materials. For example, the ultimate strength of an isotropic, defect free single crystal material in compression, is given by the density of the material multiplied by the speed of sound squared,

(2) $$S_{max} = \rho v^2_{max}.$$

Stress is equal to force per unit area or in this case also by energy per unit volume, so

(3) $$E_{max} = mv^2_{max}.$$

E_{max} is the bond energy, which is the sum of the kinetic and potential energies. The caveat about compression eliminates the crack sensitivity some materials have in tension.

In a crystalline solid, a dislocation moving under the influence of applied stress is subject to the Lorentz contraction equation. From "Mechanical Behavior of Materials"[7] we read, "the energy of a screw dislocation increases from its rest energy, ε_{so} to ε_{sv}, where

(4) $$\varepsilon_{sv} = \varepsilon_{so} (1-v^2/c^2)^{-1/2},$$

and from "Physical Metallurgy"[8], "In analogy with relativity theory, one defines,

(5) $$U' = c^2 m_D,$$

and finds the equivalent dislocation mass to be,

(6) $$m_D = m_o (1-v^2/c^2)^{-1/2},$$

where $m_o = U/c^2$".

(7) $$U' = U (1-v^2/c^2)^{-1/2},$$

U' is the dislocation total energy and U is the energy of a stationary screw dislocation. "The case of an edge dislocation presents

32

considerably greater mathematical difficulties, but the results are largely comparable".

In General Relativity, Einstein generated an equation that deals with the elastic properties of "empty space". Interestingly that tensor equation, $T_k = C_{hk}S_h$ is identical to the general tensor expression for stresses produced by piezoelectrically induced strains in a piezoelectric crystal.

Quantum effects have to do with mean values and probabilities as does the kinetic theory of gases. Uncertainty is implicit in the kinetic theory and an uncertainty principle can be derived from that simple classical theory. For many years quantum theory has required that particles fill all space, winking in and out of existence constantly. Photons certainly fill space with a density of about $1 \times 10^8/m^3$. A Higgs fluid filling space has been proposed to account for particle mass. No one seems to have examined the properties of such a fluid or gas as a particle ensemble however. In 1978 when I applied the kinetic theory of gases to such a proposed particle ensemble, an equation looking like the Heisenberg Uncertainly Principle readily appeared before my astonished eyes. Excitedly, I wrote up the consequences of my discovery and presented them to a professor whose PhD was in Physical Chemistry, I believe. At the end of the presentation, I concluded that order often arises out of chaos. His only comment or at least the only one I remember was that he could not agree with my last statement. Not yet discouraged, I wrote a complete article showing how no fact of physics was violated by assuming that space was filled with sub-electron particles and submitted it to Speculations in Science and Technology. The paper was rejected with a single comment, mentioned earlier; if space were filled with particles they would have to be bosons not fermions.

Following sections demonstrate how and why special relativity and quantum mechanics follow logically from xion material properties and the kinetic theory. They will show how quantization arises.

Motion through a material is resisted by its molecules. The same is true for a xion material. As a result, the Lorentz transformation equations have an obvious physical cause, an explanation lacking in Special Relativity, where the only explanation is that the speed of light is constant, for some reason. The

33

knowledge that motion is relative to the universal xion material finally resolves the "Twin Paradox". In the section Lorentz Factors and the Twin Paradox that follows, the author shows why the current accepted solution to the paradox is flawed.

Lorentz Equations and the Twin Paradox

Given a solid material of length, d, moving at velocity v, in a xion material, the atoms of the material will be opposed by the force of collective xion scattering. Compton showed that x-ray photons transferred energy to electrons during collisions between the two different particles. The force on the moving body is given by the change in momentum due to particle collisions,

(8) $$F = \Delta(Mv)$$

One can visualize the relationship between the velocity of a metal beam moving in a gas and the resistance force, by imagining the beam, length d, spinning about one end with a tangential velocity, v. The centripetal force on the beam would be,

(9) $$F = Ma = Mv^2/d.$$

That force would cause an elastic contraction of the beam, $F = \alpha\Delta d$. Thus, the stress,

(10) $$S = F/A = (\alpha d/A)(\Delta d/d) = Y\varepsilon ,$$

$$Y = Mv^2/\Delta dA \quad \text{and} \quad \varepsilon = \Delta d/d ,$$

From the Lorentz contraction, $\quad \Delta d = d \left[(1-v^2/c^2)^{1/2} - 1\right]$

Also from the Lorentz contraction, we know that $d = d_o (1-v^2/c^2)$. The greater v is, the smaller d is, and the greater the force on the bar. The time dilation equation is obtained from the contraction equation by noting that $v = d/t = d_o/t_o$, so,

11) $$t = t_o (1-v^2/c^2)^{-1/2}.$$

The time dilation equation produces that interesting paradox involving twins, in uniform motion relative to each other. Let us review the problem again. Suppose one twin stays home while the other takes a spacecraft journey. Each twin, in uniform motion, observes that the other is aging more slowly than himself, based on identical clocks they observe in each other's possession. Clearly, this is impossible. In an effort to resolve this puzzle, which casts doubt on the time dilation equation itself, some physicists have proposed that the twin who leaves the earth ages more slowly, because he experiences accelerations when he leaves, turns around and slows to land. These accelerations slow his clock relative to the stay-at-home twin. But appealing to accelerations takes the problem out of special relativity context in which it was conceived. More serious perhaps, is the fact that those who propose this solution conveniently forget the gravitational acceleration experienced by the earth bound twin. Einstein showed that inertial and gravitational accelerations are equivalent. One can arrange, in thought at least, for both twins to experience the same acceleration forces over the period of their relative motion by adjusting the spaceship propulsion and orientation to always provide exactly one g of continuous acceleration. As a result, one is back to the original paradox.

Since the traveling twin does indeed experience the resisting force of the xion gas in which he is moving with high velocity, resolution of the problem is straight forward. It is the twin in motion relative to the xion material that is aging more slowly, provided his net acceleration is greater than his brother's. On Earth, a clock is traveling at about one thousand miles per hour, relative to the xion gas, due to Earth's spin. If one puts an identical clock in orbit with a velocity of seventeen thousand miles per hour with respect to a fixed point on the earth, the time difference after n orbits will be given using the time dilation equation, but with the velocity difference between the earth clock and the orbiting clock., not the orbital velocity.

$$(12) \qquad\qquad t_o = t_e \left(1 - v_d^2 / c^2\right)^{-1/2} ,$$

t_o is the elapsed time on the orbital clock, t_e is the elapsed time on the earth clock, and v_d is the velocity difference. Remember, v^2 reflects acceleration effects of circular motion as well as the linear resistance to motion.

Perhaps it would be useful here to review the text book analysis of the Michelson-Morely experiment which gave rise to the Lorentz transformation equations. The analogy often used is a boat on a stream. The stream velocity represents the ether wind generated by the rotating earth. The boat represents a photon traveling through the ether. This is not a satisfactory analogy because the water offers resistance to the boat which must be constantly accelerated in order to overcome that resistance. The boat is more like the interferometer which encounters resistance from the ether. A poor analogy is worse than none because it introduces confusion. If we simply let c be the speed of light and v the speed of the ether wind, we find the time necessary for a photon to go a distance ,d, against the wind and return the same distance with the wind is,

(13) $T_x = [d/(c-v)] + [d/(c+v)] = d[(c+v+c-v)/(c^2-v^2) = 2dc/(c^2-v^2)$.

The time for a photon to cross the wind a distance, d, perpendicular to the wind and return is,

(14) $T_y = 2d/(c^2-v^2)^{1/2}$.

Substituting numbers for the various quantities quickly shows T_y is less than T_x. But the MM experiment showed that the times were in fact identical. Lorentz concluded that the interferometer arm in the direction of motion contracted so that d_x was not equal to d_y. In particular, he showed that,

(15) $d_x = d_y(1-v^2/c^2)^{1/2}$.

Now one can employ the boat on a stream to understand this result. If a boat tries to move through the water at some speed, v, it will encounter a resistance. Force must then be applied to keep the boat moving at a given velocity. Therefore, since $a = v^2/r$, we see that the velocity squared term implies acceleration. As one tries to move faster, the resistance increases. Here the analogy fails again because a boat can move faster than the speed of sound in the water, but nothing can travel faster than the speed of light in a vacuum. If the boat were composed of a cylinder of thin aluminum filled with a balloon full of water and the propulsion system took in water, heated it , and forced it out in a jet, the boat would break up before it reached the speed of sound in water, at the water's temperature and pressure. This more closely represents the interferometer moving in vacuum.

The term,

(16) $$(1-v^2/c^2)^{1/2} - 1,$$

is the strain in the material under pressure. In the Lorentz contraction, for instance,

(17) $$1 = l_o(1-v^2/c^2)^{1/2}, \quad \text{so } \varepsilon = (l-l_o)/l_o = (1-v^2/c^2)^{1/2} - 1$$

which in turn is equal to,

(18) $$[(c^2-v^2)^{1/2} - c]/c = \Delta c/c$$

If we set

(19) $$c^2-v^2 = s^2, \quad \text{then,} \quad (20) \qquad \varepsilon = (s-c)/c.$$

This is telling us that the xion material, is compressed and the frequency of the light wave is increased in that region. Since a clock period is the reciprocal of its frequency, an increased frequency means a slowing clock.

The evidence is clear, vacuum is a material. Referring motion to that material resolves the twin paradox. Photons also fill space and may be used as a substitute reference frame, hence, a twin's absolute motion can be determined by reference to the cosmic radiation background, CRB.

Kinetic Theory, Diffusion, And Quantum Mechanics

In order to understand such properties of a gas as temperature and pressure, statistics must be introduced. We talk about the mean free path (mfp) and the mean time between collisions (mtbc). Such concepts also apply to diffusion in a solid, but the mfp becomes the lattice distance and the mtbc becomes the time required to jump from one lattice site to an adjacent site. The ratio of the two gives the maximum speed of a stress wave in the material. For a xion material, that maximum is the speed of light. Since the speed of light is also the ratio of Planck's length [9] to Planck's time. We assume that Planck's length is the shortest distance between particles and Planck's time is the least time to travel between adjacent sites in the xion material. In any fluid, the product of the mfp and the mean particle momentum is equal to the mean action. For an xion gas

(21) $$<x><p> = <h>$$

For a solid, the equation could imply reciprocal motion; a particle vibrating about its lattice site, for example. The pendulum serves as a model for such motion.

If one wishes to express this relationship in probability wave format, we must convert the mfp to a wave length, $<\lambda> = 2\pi<x>$. Momentum stays the same because $<p> = 2\pi<x>/2\pi<t>$, so, $<x><p> = h/2\pi$. This is a familiar equation. It would look like part of the Heisenberg uncertainty relation, if h were equal to Planck's constant. We can make it even more so, if we consider $x = ax_p$ and $p = b p_p$, where x_p and p_p are Planck's length and Planck's momentum respectively, and a and b are integers, then

(22) $$x p = ax_p b p_p = nh/2\pi.$$

Here, n is a positive integer and h is Planck's constant. Now n can be 1,2,3,etc. so the expression becomes

(23) $$<x> <p> \geq h/2\pi$$

This is because n is an integer greater than or equal to one, and because action is measured in natural units of the mean action, Planck's constant, $<h> = h$. The equation is now the well known Heisenberg Uncertainty Principle. It says the product of length and momentum can be measured with a ruler whose minimum unit of measurement is the quantum of action, h. Planck's constant, like mfp and mean momentum, is an average value. As such, action, like position and momentum is reported as an integral number of average values in a quantized system.

The word quantum can be confusing because of its different meanings. In measurement, it refers to a quantity or amount. Particularly, it designates a portion or part of. I would add that it should contain the idea of a uniform part of. The confusion comes when one talks about a quantum leap, which usually implies a huge step. Quantum can mean a gross quantity, although I believe the intent is often to simply convey the notion of a discontinuous step. Because in physics quanta are very small increments of energy, one often is not sure whether the term is intended to describe a small amount, a uniform part, or a discontinuous step. Personally, I would like to see the term, "quantum leap", removed from the language, but it seems to be a favorite of journalists. For the purpose of measurement, I hope it is clear that it means a relatively small, uniform portion, which can be summed as integer units to constitute the whole. There is also the implication, when considering measurement error, that the unit or quantum of error is an average value of a normal distribution.

In wave format the mean free path or lattice distance is $\lambda = 2\pi<x>$, so equation (21) becomes, $\lambda p = h$ when the average signs are dropped, and h is equal to Planck's constant. The new equation is the de Broglie expression for matter waves, which Born correctly interpreted as probability waves.

$$(24) \qquad\qquad p = n\, h/\lambda$$

The probability wave function, so confusing when first encountered in Schrödinger's wave mechanics, is readily explained by showing its equivalence to Einstein's [10] equation for Brownian motion,

$$(25) \qquad\qquad <R^2> = (2kT/K)\, t.$$

$<R^2>$ is the mean square displacement of a particle in a fluid having thermal energy kT, K is a constant dependent on fluid viscosity and particle size, and t, the elapsed time. The root mean square displacement, $<R^2>^{1/2}$ is the expected value of distance, r for a diffusing particle relative to its initial position after a given time, t. For self diffusion in a cubic crystal lattice, $<R^2>^{1/2} = n^{1/2} <r>$, where n is the number of jumps (jump frequency multiplied by the time) and $<r>$ is the lattice parameter. Because the lattice particle is vibrating about its lattice position, the diffusing molecule carries a wave packet with it as it jumps from site to site.

A particle random walk of sufficiently many steps can be approximated by a probability wave function, which is analogous to the random walk equation. $[<R^2>]^{1/2}$. For a wave function, the probability of finding x between x and x+dx is the integral of $\psi\psi^*$ dx over that range. Since $\lambda = 2\pi<r>$, $n<r>^2 = n\lambda^2/4\pi$

(26)   the integral of of $\psi\psi^*$ dr from r to r+dr = $[n\lambda^2/4\pi^2]^{1/2}$ = $[nr^2]^{1/2}$.

Thus, if ψ = Asinωt represents a simple wave, one can see that these two approaches are equivalent. In other words, Quantum Mechanics is the Kinetic Theory of Gases applied to a very dense material of very small particles. Electrons behave as Brownian particles in a material composed of much smaller fermions. Note that when waves are employed to calculate elapsed distance, time is carried as an imaginary quantity. Much is made of the idea of imaginary time, but it is simply a mathematical technique for dealing with a fourth dimension. Tensors are an alternative technique to accomplish the same end.

By now it should be clear that a wave is a mathematical creation like the normal distribution or the average value. In this sense all waves are probability waves. Ocean waves, i.e. "real waves" are like the rainbow, a creation of the brain. The location of any particular water molecule is probabilistic. That does not make the water wave any less real to an observer, but we must realize that the wave is a subjective, virtual reality.

While we are on the subject of waves and the fact that they are all probability constructions which have in the observer's mind an emergent reality, let us reconsider the phenomenon of optical interference. Young's demonstration of this fact clinched the argument with Newton in Young's favor, only to have it overturned by Einstein's Nobel Prize winning photoelectric effect experiment. But if light were just a particle, there would be no diffusion of a beam of light passing through a pinhole, and there would certainly be no interference effect when passing through closely spaced double slits in a opaque sheet of material. But suppose the vacuum is a crystal composed of subelectron sized particles? Now, using the drunk man's walk analogy, we can readily see that photons, electrons and even neutrons would diffuse through the lattice, even if injected one at a time, to create a diffuse pattern of material on a surface behind a pinhole and an interference pattern behind a closely spaced double slit. In Quantum Mechanics this effect is attributed to the wave-particle duality. In fact no such duality exists. It doesn't make sense. What does exist is a face centered cubic crystal, perhaps dynamic, composed of unknown particles in rapid random motion. Recently, the existence of such a fabric has been demonstrated by showing that not all frequencies of electromagnetic radiation exist in the microwave background that permeates the vacuum. Specifically, frequencies not allowed in a faced cubic crystal are not found. It seems that Maxwell and Kepler may have been at least partially correct after all when they suggested space was a crystal sphere. I believe the five Platonic crystal structures can all be found in the fcc arrangement.

In the case of the double slit drunk men's walks, one must realize that it is not only the existence of a lattice that gives the particular pattern, but also the distortions introduced in that lattice by the diffusing specie. Centers of stress affect the jump direction and frequency probabilities. Although the details are complex, anyone can readily see why most of the diffusing particles wind up below the center of the distance between the two slits.

Let us now consider that peculiar quantum mechanical phenomenon, tunneling. Imagine an energy well of height E_w, containing a gas with a standardized normal distribution of energies. The average energy is $<E>$, and the variance is given by the standard deviation, σ. Let E_w represent the 4σ value of the distribution. That means about 99.995% of the gas molecules have energies less than E_w. Thus, about 0.005% of the molecules possess energies greater than or equal to E_w. If the gas density is 1×10^{20} molecules per milliliter and the well has a volume of 1×10^{-10} ml, then there are 1×10^{10} molecules in the well. Suppose the opportunity frequency per molecule per second to surmount the energy barrier is 1×10^6 per second, then, on the average, the probability of a molecule getting out of the well is $1 \times 10^{10} (1 \times 10^{-6})(0.00005) = 0.1$ per second , or six per minute. If you thought that every molecule had the average energy, i.e. $\sigma = 0$, you would conclude that no molecule could escape. Faced with the fact that six per minute did escape however, you would be forced to conclude that they mysteriously tunneled out. I expect this is exactly the situation with tunneling electrons. With a very high density of xions and a low standard deviation in their energy distribution, all electrons in the xion material appear to have essentially the same energy. In fact a few occasionally have much higher than average energies. These are the electrons that seem to tunnel through an energy barrier.

One may have noticed a problem with the tunneling model based on the idea that a few electrons actually do have energies greater than that of the barrier. In such a model the barrier thickness should not matter, in fact , it does, the thinner the barrier, the greater the probability of the electron tunneling out. Of course this makes perfect sense. We did not consider the time necessary for the electron to traverse the barrier. In an energy distribution of electrons driven by random xion vibration, a particular hot electron only exists for a given period of time, on the average. It is actually the product of the electron's energy and its lifetime at that energy that is important in my tunneling model. This product is called action. Since the thinner barrier requires less time to traverse than a thicker one, the probability of tunneling increases as the barrier thickness decreases. If we express action in terms of Planck's constant, then the natural logarithm of the probability of tunneling through a particular barrier should be proportional to the ratio of the action to Planck's constant. The natural log of the probability will also be proportional to the ratio of the electron energy to the barrier height. Thus, the probability of jumping the barrier will be proportional to $e^{Et/h}$. This simple model looks very similar to the quantum mechanical tunneling model.

The same phenomenon accounts for radioactive decay which is itself a tunneling effect. Since the xion supply is virtually unlimited, one can compute a half life for the decaying specie. In the case of the gas in a well example above, at six per minute, the half life would be about 1,585 years. The calculation assumes the molecular energy distribution does not change as a result of the loss, which in this case is unrealistic, but is not unrealistic for an xion material because of it enormous density.

Throughout this section, I have speculated about a particle based material composing the vacuum. For some ancient Greek thinkers, the ether consisted of the finest atoms. Certainly, they thought, if something occupied space, it must be like a gas. Think about the virtual particles of quantum mechanics popping in and out of existence, or the photon gas that is the cosmic radiation background. But the luminiferous ether of Maxwell must be like a solid in order to support electromagnetic waves. This conflict of properties is one of the reasons Lorentz's and FitzGerald's support for the ether was rejected. How could a material be a rigid solid and still allow the earth to pass rapidly through it on its journey around the sun?

Attempting to solve this puzzle, the author came up with the idea of a dynamic crystal. Normal definitions for the three primary phases of a material are incomplete, because they usually ignore time. When one enters water slowly, the fluid readily flows around him, offering little resistance. When the same liquid is struck at 200 miles per hour, it feels like a rock and causes similar damage. Even air struck at very high speed would feel like a solid. Therefore, the rigidity of a material depends on the speed of impact relative to the speed of a stress wave in the material. Let us address the behavior of a particle assembly, in the gas phase, at times short compared to the mean time between collisions. In particular we are interested in a high particle density gas composed of fermions, such as neutrinos. Since they could be particles which have been undetected thus far or some other presently unsuspected particle, I have called them xions

Consider a monatomic gas at standard conditions of temperature and pressure. Take a mental snapshot of the gas in a time short with respect to the mean time between collisions, (mtbc). You will "see" that most atoms are a mean free path apart. This looks like a lattice. Some of course are not and appear at various distances from lattice positions. Vacancies or interstitial defects also appear occasionally. Take another picture one mtbc later. Again the particles seem to be arranged in a face centered cubic lattice, but when the two photographs are aligned according to the lattice atoms, some particles have moved relative to the lattice position and the vacancy and interstitial defects may have moved as well. If we now make a mental motion picture of the gas at one frame per mtbc and register each frame to the same reference coordinates, we will see the defects moving through the lattice as lattice particles vibrate. This is a dynamic crystal.

Make our material a dense xion gas, with the mean free path or lattice parameter equal to Planck's length. The speed of light, (in which c equals $<x>/<t>$, the mean free path divided by the mean time between collisions, also equals L_p/t_p, where L_p is Planck's length and t_p is Planck's time. Now consider the equation for the force between two masses due to gravitational attraction, $F = G m^2/r^2$, as r is decreased, the force between the particles is increased. At Planck's length, 1.61×10^{-33} cm, the force between adjacent particles of even miniscule masses can be very large. For the electron with a mass of 9.1×10^{-28} g, the force between two of them separated by Planck's length would be 2.1×10^7 dynes. For comparison, recall that a dyne is defined as the force between two one gram masses separated by one centimeter. With such a strong force one can appreciate that even a dynamic lattice can sustain shears forces. This fact allows a xion gas to behave like a rigid solid, transmitting shear waves at the speed of light while still permiting particles to diffuse through it at a rapid rate with only a small resistance. This is of course speculative, since we have no estimate for a xion mass.

Most particle-particle collisions are not direct hits; therefore the particles not only exchange linear momentum, but also cause and exchange rotational momentum. A virtual or quantized particle, having a mean value of some parameter averaged over time, representing a real particle, will have quantized angular momentum. In the case of a virtual or quantized xion in a quantized lattice, this angular momentum could be called spin.

Interstitials and vacancies are mechanical stress centers. Vacancies are centers of tension, while interstitials are centers of compression. Being a dynamic crystal, lattice site occupation changes every mtbc, but on the average the same number of particles is missing from lattice sites. It appears then as if a vacancy were a real particle that simply moves from one lattice site to another. Similarly, the interstitial appears to be the same particle moving around in the lattice. The spin of the quantized xions will be transferred to the virtual particles, thus vacancies and interstitials in the dynamic lattice possess quantized angular momentum or spin.

For a xion gas, the dynamic crystal has a very fine lattice. Vacancies, as centers of (tensile) stress could be considered the origin of positive electric fields. Interstitials, as centers of negative (compressive stress) would then be responsible for negative E fields. The electric field is therefore, a mechanical stress field. The magnetic field is the lattice reaction to a change in the stress and therefore has no separate or different particles associated with it, only orientation. Application of an electric field to a dynamic xion gas crystal causes the defect particles to move through the interaction of the stresses.

Protons, electrons and neutrons would be emergent particles. Identification of various particles with lattice defects would make a good parlor game for particle and solid state physicists.

An interesting consideration has recently come to light relative to the idea of an xion lattice. The structure would be face-centered cubic. Perhaps you have heard of the five Platonic solids, which Kepler thought constituted the solar system. Each of these "solids" can be found in a face-centered cubic lattice. In the book, 'Descartes' "Secret Notebook" [11], page 239, the author says,"Jeffery Weeks exposed in an article in the Notices of the American Mathematical Society, a theory that showed that tetrahedral, octahedral, and dodecahedral models of the geometry of the universe agree very well with the new findings". Those findings are from the latest data taken on the cosmic radiation background, which show that certain frequencies are missing in the microwave background radiation the permeates the universe. In other words, the universe looks as if it were some kind of face-centered cubic crystal.

The dynamic crystal was a desperate search to avoid a seemingly unacceptable conclusion – vacuum is a solid. To even support the ether theory automatically invites raised eyebrows, smug sneers, and a label, zealous crank. To further claim the ether is a solid invites immediate dismissal. But nature behaves as it does and we must accommodate to it. Something fills the vacuum, a gas seems unlikely. The dynamic crystal tries to avoid the inevitable, but it may not support light waves. Therefore, one is led to the conclusion that vacuum may be a solid and motion through it takes place by diffusion.

To summarize thus far:

1. The xion based material that fills all space can be treated like any material. Applying ideas from the Kinetic Theory of Gases, diffusion, and periodic motion generates the Heisenberg Uncertainty Principle and explains quantization and probability waves. A close look suggests the material may actually be a solid. The Heisenberg relation then looks like it is referring to the vibration of a xion particle about its lattice position.
2. The Lorentz transformation equations result from the pressure experienced by a body moving in the xion material..
3. The universal xion material is the absolute reference frame that allows resolution of the Twin Paradox. The CRB may be substituted for this purpose.
4. Tunneling can be explained by a distribution of energies with a variance less than the measurement error.

Electron Orbits and Simple Harmonic Motion

Electron motion in orbit about the proton in a hydrogen atom is the result of Brownian motion and the attraction between unlike charged particles. Each path between collisions is like a spring executing a cycle from its contracted state to its extended state and back. In a line, such motions are passed along as radiation. But if the line closes on itself, the radiation is contained. This makes sense from a vibrating particle point of view. If the vibration completes a cycle perfectly, energy is conserved. If the cycle is not completed, some energy has been lost to the lattice in the form of a wave transmitted throughout the lattice. That is what Bohr found for the electron orbits. An integral number of mean free paths or wave lengths are required to close an orbit exactly. The closed loop does not have to be a circle or even an ellipse as long it contains an integral number of wave lengths. The spectral orbits are the most probable path values for the electron given the energies in the Brownian motion equation which has been quantized.

While we are on the subject of waves, now is a good time to explain Born's famous equation, as the author promised.

(27) $$qp - pq = h/2\pi i \ .$$

In the xion lattice, xions vibrate about their lattice positions, reflecting their actual motion between collisions. There will be an energy distribution of these oscillations between zero and m_pc^2. Each oscillation behaves like a spring or a pendulum in a gravitational field. The total energy of a mass in uniform circular motion is $E = mv^2$, where v is the tangential velocity. Recall that ma $= mrv^2$ in this case where a is the centripetal acceleration and r is the radius of the circle. If the velocity of the oscillator reaches the maximum velocity for a wave in the material in which the oscillator exists, as in a single crystal, then $E = mc^2$. The total energy is the sum of the kinetic and potential contributions. The maximum velocity equals the ratio of the mean displacement, x, to the mean time to travel that distance, t. In a gas that is the ratio of the mean free path to the mean time between collisions. Using this information we can modify the total energy equation to read, $E =$

(mc)(x/t). Multipling E by t gives the action, H and mc = the momentum,p so the expression now becomes, x p = H. But t is equal to the period of oscillation, τ, divided by 2π, thus, $E(\tau) = x\ p = h/2\pi$, or in wave format, $\lambda\ p = h$, and $E = hf$.

In four dimensional notation, time is expressed as an imaginary number, so

(28) $E\ (it) = H = x_n\ p_n$ and $E(\tau) = x_n\ p_n = h/2\pi i$

One should realize that x p, where n = 1, is actually the sum of two portions of a cycle,

(29) $(x'/2)p + (x'/2)p = x\ p = h/2\pi i$

But, if we let q equal x'/2, and q p= -p q this expression can be written as,

(30) $q\ p - p\ q = h/2\pi i$

Readers will recognize this as the Born equation, number (27) above, which as mentioned previously is engraved on his headstone. So, you see this famous expression of Quantum Mechanics is simply another way of writing the equation of simple harmonic motion for a particle tied to a lattice position in a cubic lattice In particular, it makes the connection between the oscillation of a particle in its lattice, and the gravity or electromagnetic wave transmitted by the lattice. Vibration of lattice xions would be transmitted as a gravity wave, while motion of the mechanical stress field associated with an electrical charge gives rise to the electro-magnetic wave. From this fundamental equation, one can see the origin of the energy expression for photons, $E = hv$, the uncertainty relation, $\Delta q\ \Delta p \geq h/2\pi$, and the de Broglie relation, $2\pi qp = h$ or $p = h/\lambda$, as discussed earlier.

Rest mass, Gravity and Gravitation

While speculative, our model, thus far, is grounded in reasonable expectations from the requirement that the universe be connected (Stapp modification to the Bell inequality) and analogies with other known materials. A problem arises however, when one inquires as to where the xions came from, and what fills the spaces between them. Possible answers are, they always were, and nothing fills those spaces. Another answer is that there exists an even more fundamental material, ylem, from which the xions were born.

I have suggested that xions have mass and that mass arises from the bombardment of a particle by countless other particles, causing hydrostatic pressure about the particle in question. Its response is elastic compression. The energy contained in that compression is equal to the mass of the particle, multiplied by the square of the maximum velocity of a stress wave in the medium providing the hydrostatic pressure. Recall that v^2 implies acceleration. Consider a spherical particle being constantly bombarded by identical particles. If the collisions are severe and rapid enough so that the particle in question cannot relax the distortion induced by the bombardment, it will exhibit a permanent strain equal to the imposed hydrostatic stress divided by a bulk elastic constant. The imposed stress is the energy per volume of the particle which is equal to the elastic constant multiplied by the volume change divided by the volume. Thus,

$$(31) \qquad\qquad E = mc^2 = B\,\Delta V$$

If we consider the elastic constant to be c^2, then the rest mass of the particle is equal to the change in volume due to the hydrostatic stress. Since the total stress on adjacent particles is less than that for separated particles, energy can be minimized by getting together. This appears as an attraction between particles which are held apart by the constant bombardment.

This fact generates the mechanism that holds the xion lattice together. Defect structures, in turn, distort the xion lattice, generating electic fields. Mass of a defect structure is virtual mass, like the defect particle itself. The structure is composed entirely of xions in a particular configuration. The whole idea is speculative, but it is bounded speculation and in that sense no different than Faraday's fields or various string theories.

For now, we must leave this speculation and focus on the consequences of defects within the xion lattice and what we do know about gravitation. In this model, constant bombardment of an individual particle, a fermion, generates the rest mass of that particle. If two fermions appear in the material, they would occupy less space if they were together instead of separated; therefore they attract one another. If they are charged particles, like defects in the xion lattice, in addition to gravitational attraction they will attract, if different, but repel if the same, due to mechanical stress interactions. Should they be neutral however, like xions, they will attract, unaffected by electric fields. The force of attraction is proportional to the masses of the particles and inversely proportional to the distance between them squared. This is the force of gravity. The constant of proportionality is the gravitational constant.

The definition of Planck's Length is,

$$(32) \qquad\qquad L_p^2 = hG/2\pi c^3.$$

Planck's constant is dimensionally mass times length times the speed of light, Mdc. Therefore,

$$(33) \qquad\qquad G = 2\pi c^3 L_p^2/Mdc.$$

The force of gravity per unit area, A, can be set equal to an elastic modulus multiplied by the strain. Thus,

$$(34) \qquad\qquad F/A = Gm_1 m_2/r^2 A = Y\,\varepsilon.$$

$$\text{Let } m_1 = \alpha M, \ m_2 = \beta M, \ r = \gamma L_p, \ \text{and,} \ d = \delta r$$

$$(35) \qquad\qquad F = (\alpha\beta/\gamma^2)GM^2/L_p^2 = Y\varepsilon.$$

So,

$$(36) \qquad\qquad Y = (\alpha\beta/\gamma^2)GM/L_p, \ \text{and } \varepsilon = M/L_p.$$

Since $\varepsilon = \Delta r/r$, $r = L_p$, and, $M = \Delta L_p$, where ΔL_p is the change in the effective radius of the xion due to xion collisions, which in turn is equal to m_p, setting $\kappa = \alpha\beta/\gamma^2$, and substituting in equation (36) gives,

$$(37) \qquad\qquad Y = \kappa Gm_p/L_p, \ \text{and } \varepsilon = m_p/L_p.$$

Substituting $m_p = M$ and $\delta r = d$ into the modified dimensional equivalent of G, equation (33), gives

(38) $$G = 2\pi c^3 L_p^2/\delta m_p rc = 2\pi c^2 L_p^2/\delta m_p r.$$

Since $L_p = \mu \Delta L_p = \mu m_p$, and $\rho L_p = r$, substitution in (38) gives,

(39) $$Y_G = \eta c^2, \quad \text{where } \eta = 2\pi \kappa \mu/\delta \rho, \text{ a number}$$

Now, multiply F/A in (23) by x/x , substitute (39) for Y and $m_p L_p$ for ε, to give,

(40) $$Fx/Ax = \eta c^2 m_p/AL_p = \eta m_p c^2/V = \eta E_p/V.$$

The area, A, multiplied by $r = nL_p = nAL_p = V$, the volume, (n is some integer) so

(41) $$E_p = m_p c^2.$$

As mentioned before, the rest mass, ηm_p, is proportional to the change in the effective fermion diameter, Δx, which is caused particle bombardment ; m_p is the xion rest mass. The elastic modulus divided by the strain is proportional to the gravitational constant. Of course, the particle is in a sense, fictitious, being a defect in the xion lattice. We must think of the particle and its properties as mathematical constructions for our convenience, just as the wave nature of the xion is such a construction, representing the distribution of actual random vibrations of xions in the lattice.

Returning to the separated particles, $F = Y \Delta r/r$ implies that the strain just shortens the separation distance. Unless something interferes, the process will not stop until the two masses are together. This causes a partial shielding of each particle by the other from xion bombardment, which reduces the total mass of the combined particles. Thus, when m_1 and m_2 are together, their combined masses should be slightly less than the sum of their individual masses. This looks like a binding energy,

(42) $$M = m_1 + m_2 - e.$$

For a neutron, we have a similar situation, namely

(43) $$p^+ + e^- = n + \mu,$$

where the neutrino, μ, carries away the extra mass (energy) when a neutron is formed.

Since G is the product of the elastic modulus and the strain, perhaps the reciprocal of the permittivity and the reciprocal of the permeability serve similar functions for the electric and magnetic fields, respectively.

(44) $\qquad\qquad S = F/A = e^2/4\pi\varepsilon_0 r^2 A = Y_e\,\Delta r_e/r,$

so $e = \Delta r_e$, and

(45) $\qquad\qquad Y_e = \Delta r_e/4\pi\varepsilon_0 rA.$

Thus, the Δr associated with e is much greater than the Δr associated with m, relative to the force generated by the particle's charges and masses, for the same separation distance. The electric charge is due to stress generated by a defect configuration, while the mass of the electron is due to the resistance to its motion in the dynamic lattice. Both may be the result of mechanical stress. In this sense mass could be considered charge. A unit mass charge, *, is associated with an average mechanical stress on a xion due to random particle collisions. Mass charges always attract.

(46) $\qquad\qquad Y_e = \varepsilon_e/4\pi\varepsilon_0 A,$

because, $\varepsilon_e = \Delta r_e/r.$

From the expression, $c^2 = 1/\varepsilon_0\mu_0, \quad \varepsilon_0 = 1/c^2\mu_0,$
so

(47) $\qquad\qquad Y_e = \varepsilon_e/4\pi\mu_0 c^2 A$

Similarly,

(48) $\qquad\qquad Y_M = \varepsilon_M/4\pi\varepsilon_0 c^2 A.$

Therefore,

(49) $\qquad\qquad Y_e/Y_M = \varepsilon_e\varepsilon_0/\varepsilon_M\mu_0.$

Since $\varepsilon_e = \varepsilon_M,$

(50) $\qquad\qquad Y_e/Y_M = \varepsilon_0/\mu_0 .$

Recall from page 52, equation (39), $Y_G = \eta c^2 = \eta/\varepsilon_0\mu_0,$ by substitution from (50)

(51) $Y_G = \eta Y_c / Y_M$.

Recall that G is proportional to $1/\varepsilon_0 \mu_0$.

      From General Relativity we know that mass causes light to bend in its vicinity as light bends when passing near the Sun. Displaced xions are actually responsible the curvature, because the lattice curves. Xions form a density gradient, in this model, that surrounds and permeates the Sun or any massive body. It is the density gradient that is responsible for the Sun's gravity and for the curvature in an electromagnetic wave's path. Any concentration variation in xion density will then cause distortion in an EM wave's path.

Gravity and Electromagnetic Waves

Vibrating xions in the dynamic xion crystal create regions of tension and compression. Since the system is quantized in terms of the lattice distance and vibration period, the stresses are also quantized in terms of tension and compression configurations. We associate charge with these quantized stresses. An electron may be represented as an interstitial xion in the lattice and is a center of compression to which we assign a negative charge. When an interstitial moves out of its site, the electric field must collapse at that site and the cell changes shape. This change is associated with the creation of a magnetic field. Let us identify a unit cell in this face centered cubic lattice. When the lattice particles vibrate they cause the cell to change shape. When the cell is compressed in the y direction, it is extended in the x and z directions. Compression of a particular magnitude is assigned a unit charge which is the unit electric field. When the stress relaxes, the field decreases as the cell recoils like a spring. But it overshoots and goes into tension, elongating the cell in the y direction. The compression is now in the x and z directions. This change is again associated with the creation of a magnetic field. Now the unit cell returns to its original position and the electric field is created once more at the expense of the magnetic field. What we have is like a set of connected springs.

The vibration of one cell is transmitted to the adjacent cells. Examine the spring located along the x axis. As it oscillates, a wave of oscillations is transmitted in the x and –x directions. When the x and z direction springs are extended, the y spring is compressed. The ends of the extended springs are compressed; the end of the compressed spring is extended. When they relax, they all go to their opposite stress configurations, so three waves are generated, one in each direction. Looking at the wave propagating in the x directions, + and -, you see the compressions and tensions in the y and z directions alternating with each other, $90°$ out of phase. If we choose the y direction as the electric field direction, the z direction will become the magnetic field direction. Our wave looks like an electric field vector decreasing as a magnetic field vector increases, such that when one is zero, the other has reached its maximum value. Then the direction changes and the maximum decreases to zero, while the complimentary vector increases to its maximum value again, but in the opposite direction. Thus, the lattice generates electromagnetic waves in all directions. The waves should have a black body radiation distribution for the average temperature of the material.

One can see from this model that String Theory may be just another way of looking at the xion crystal.

To reiterate, a xion in simple harmonic motion is the smallest oscillator. Lattice oscillations are then composites of many xion oscillations. Using the Maxwell- Boltzmann distribution of relative particle distances, we can determine from the variance what fraction of the population is at a given distance from the mean at any instant of time. As time progresses in units of the vibration period, a given average site will seem to show a xion vibrating in a complex, random pattern. Through Fourier analysis, this average site will show a vast array of simple harmonic motions of different frequencies and amplitudes. Because the end points are fixed, the frequencies must be quantized.

The unit wave is like a spring that goes from compression to extension and returns. It is represented by the formula,
(53) $qp - pq = h/2\pi i$.
Lattice waves can have any frequency, as long as they obey the equation, $E = hf$, and $f = cL/n$, where c is the speed of light, L is the length of the crystal, and n is a positive integer less than L divided by Planck's length.   If the waves are associated with electric charges, they are electromagnetic waves.

Conclusions:

The notion of a xion crystal is interesting in and of itself and it seems to solve some fundamental problems. For example, why are all electrons alike and what accounts for radioactive decay.

Four additional consequences of the xion crystal model of the Universe are;

1. Because the density, pressure, and temperature of the xion material may vary with time, the values of the speed of light, the fine structure constant, and the gravitational constant may have been different in the past. Recall that wave velocity squared is proportional to the ratio of the pressure to the density, $v = P/\rho$. If the speed of light were less two billion years ago, as suggested by Lamoreaux and Torgerson[12], it is likely that it had an even smaller

value further back in time. The universe in the distant past would then have appeared smaller than it actually is, and the rate of universal expansion would have appeared to be slower that it actually is. Therefore, the rate of expansion would appear to be increasing with time, contrary to expectations. That is exactly what has been reported recently.

2. Dark energy and dark matter may be manifestations of the xion material.

3. The gravitational constant is proportional to the square of the speed of light. As such, G also may not be a true constant, but change with density, as would the speed of electromagnetic waves.

4. Xion vibrations in the lattice generate gravity waves, while the motion of mechanical stress fields associated with electric charges generate electromagnetic waves.

Neutrinos, Xion Rest Mass, Antiparticles, and Quarks

Thus far we have made the case for an all pervasive xion material. Acknowledging such a material and applying the Kinetic Theory of Gases resulted in the demonstration that Quantum Mechanics is a mathematical way to deal with the random motion of particles. The presence of the material in the universe was shown to account for the Lorentz transformation equations and resolve the Twin Paradox of Special Relativity. With the concept of a xion crystal, the author was able to extend the remarkable results of the model to suggest an explanation for the origin of protons, electrons and neutrons. As surprising is the explanation of rest mass and gravity the model affords.

Discussion of neutrinos, xion mass, antiparticles, and quarks was left for later, because the implications of what was shown to result from the model were already overwhelming. Too much information can be confusing. The neutrino could be the xion itself. The neutrino possesses a small rest mass. If the neutrino density in space is as large as that of the xion material, the consequences are dramatic, changing the present view of the origin of the universe. Finally, we will address the subject of antiparticles and quarks.

When a neutron decays, an interstitial xion has left the vicinity of a vacancy in the xion crystal. More familiarly, an electron has been emitted from a neutron leaving a proton. But some additional energy (mass) is left over that escapes as a neutrino. Clearly, the total energy (mass) of a neutron is less than the sum of the separate proton and electron masses. That energy is the result of a distortion of the dynamic lattice due to an extra xion. Since, the sum of the energies of the separated proton and electron are more than that of the neutron, when the neutron decays, the energy associated with the difference translates into a particular amount of mass (that of the neutrino) multiplied by the speed of light squared. This can also be seen as the result of the extra xion causing a chain reaction in the lattice which appears to be a xion moving through the lattice at the speed of light.

The neutrino possesses rest mass, as has been experimentally verified, and the xion should also. Xion bombardment generates the rest mass of a xion, so, now we have the same problem as with the neutrino, that is, a particle with positive rest mass cannot travel at the speed of light. The author has already explained that a xion is a particle, not a wave. There is no wave-particle duality. A wave packet may or may not be associated with a particle, but the two things are not the same. Think about air. Molecules can travel only a short distance, a mean free path (mfp), before a collision. The mfp divided by the mean time between collisions (mtbc), is the speed of sound. Because the molecules are constantly bumping, they can transfer information or energy at sonic speed. Phonons are wave packets generated by the random vibration of molecules in a solid. The same model applies to a xion crystal Their average speed is that of light. That means they can convey information or energy at the speed of light. It does not mean that a single xion travels for any distance at that speed. Each xion is executing a random vibration. Its diffusion path is Brownian in nature in the absence of any bias.

The resistance to motion which Lorentz assumed in his contraction equation and which we see in the slowing of the earth's angular momentum and Pioneer Ten and Eleven's velocities, is also effecting the motion of solar systems and galaxies. Take a star of mass, m at a distance, R from the center of a galaxy of mass, M. The centripetal acceleration of the star is M/R^2, so the star's orbital velocity should be, $v = (M/mR^2)^{1/2}$. But suppose there is a resistance to that motion which is proportional to the velocity, then $v_0 = (M/mR^2)^{1/2} - nv_0$. The result is the actual measured velocity is $v_0 + nv_0$ or $v_0(1+n)$, while the expected velocity is v_0. Suppose $n = 0.2$, the measured velocity is 20% greater than expected. Since one can't change the mass of the galaxy, the conclusion is that there is extra mass in the galaxy that cannot be seen. That mass is assumed to be due to "dark matter" that resides in the vicinity of the galaxy. Dark energy, sometimes called, quintessence, would just be the energy of the xion material itself.

If xions possess mass, the mass of the universe may well exceed that necessary to stop universal expansion and turn it into universal contraction at some point in the future. That would mean we live in an oscillating universe. Gone is the singularity. Big Bang sets off an expansion phase; Big Crunch ends the contraction phase as we enter another cycle. Interestingly, the ancient Sanskrit speaking people of India, who believed in five states of matter, the fifth being that which fills space, may have correctly guessed that nature is an endless series of cycles.

Defects, particular xion configurations in the xion lattice, such as protons, electrons, and neutrons, all have their antiparticles. What are they? Consider how mass is created by bombardment. In the dynamic crystal model, the energy of bombardment is transferred to rest mass, which is the compression of the particle. This is like the compression of a spring. In addition, the xions making up the lattice vibrate about their lattice sites. These vibrations also resemble springs. Kinetic energy is expended or work is done to compress the spring, which now has potential energy. If the work done is positive, the acquired potential energy is negative. But suppose the work is done to extend the spring; is the acquired potential energy still negative? We are having a problem here trying to use one sign convention for two different things. We need another convention. If we call the spring compressed a distance, r, a particle, E, of potential energy, ξ, then the spring extended the same distance will be an antiparticle, E*, with potential energy, ξ, as well. When we take the center of the spring and move it a distance, r, we will have created a particle pair, each having the same potential energy. In a crystal lattice, when a particle vibrates it generates a tensile force on one side and a compressive force on the other, relative to its particle neighbors, fixed to the lattice.

Suppose the vibrating particle moves one half a lattice distance. Two particles are created in this process, one a center of compression, the other a center of tension in the lattice. We could label these stress fields minus and plus, but now they would be a xion - antixion pair. These particles would only exist momentarily as they indeed do in vacuum. So we see that particle pairs may be simply configurations the vibrating lattice generates. This sounds like String Theory. In that theory, particular vibrations are associated with particular particles. String theorists talk about vibrating membranes, the author talks about a dynamic xion lattice. Chances are we are talking about similar things, a mathematical construct that allows one to explain the origin of particles and fields in vacuum filled with xions.

Suppose the vibrating particle moves half a lattice distance, a vacancy and interstitial have been created simultaneously, but the arrangement is unstable and the xions immediately relax into their former positions. This looks like particle creation and annihilation. The movement of a xion from a lattice position to an interstitial site, is another xion-antixion configuration, which is also unstable. While the xion in a lattice position has no charge, the vacancy can viewed as a positive charge, and the interstitial as a negative charge. One might consider the interstitial an electron, but the proton would probably have to be a more complex configuration. Even the idea of an electron as a interstitial is a problem because of the vast difference in the diameter of an electron and the size of the xion lattice interstitial site. If the size of the electron were to include its electric field (the mechanical lattice distortion), that could account for the size difference

Protons are composed of quarks, but what are quarks? In the dynamic crystal model, quarks may be quantized stresses. There are three principle stresses, so there are three principle quarks, red, green, and blue. There are two shear stresses for each principle stress. Since each stress has two possible directions, there are 6 + 12 = 18 quarks associated with those stresses - red , green, blue, up – down, strange – charm, top – bottom, and their antiparticles Then we have couples, responsible for spin. There are two couples for each principle stress. Given two directions for each principle stress, that equals 12 couples. Therefore, there are a total of 30 quantized stresses and couples as required by electro-chromo dynamics. If the model is valid, it is clear why no quark can be isolated, disembodied stresses do not exist.

Xion Crystal Model; Conclusions, Implications, and More Speculations

The neutrino has been found to possess rest mass. Xions too should have a small rest mass. If they do, they are a form of dark matter and dark energy is just the energy of the xion vibrations. The little mass xions possess could easily push the total mass of the universe over the necessary value required to stop universal expansion and reveal the system to be a magnificent oscillator.

Antiparticles turn out to have a simple explanation. Their creation and annihilation is readily understood, but their scarcity is an enigma. Similarly, quarks have a simple explanation, although their analysis in particle collisions is certainly complex.

Now we have insight about how String theory relates to the xion lattice. Strings are the vibration patterns in the dynamic lattice that define particles and their interactions.

Over the years mathematicians and physicists have pondered why mathematics is able to model physical behavior. It is clear that all mathematics has a very humble beginning, it was built from nothing. From zero one can obtain +1 and -1. With these three quantities alone, all mathematics is constructed. In physics, we see a similar situation; from the xion crystal, which has no net charge, come the vacancy, with a charge of +1 and the interstitial, with a charge of -1. These defects could be protons, (but positrons make more sense) and electrons. Together with neutrons these particles form atoms, the building blocks of all material, except the xion material itself. Is it any wonder that mathematics models physics as well as it does?

Thermodynamics has provided us with an intriguing question, one that was the subject of a particularly interesting science fiction story. In the story, the question "How can the second law of thermodynamics be reversed" was solved in an unlikely manner. The intergalactic computer, built to find out, demanded more and more circuits. The civilization was dying. Finally, only one scientist was left. He interrogated it one last time before he died. The giant computer hummed and blinked the proclaimed, "Let there be light". If the reader accepts the premise of this booklet, there may be a more obvious way to turn things around. That is the subject of the next section on entropy. Given an all pervasive xion material in which the xions possess rest mass, perhaps the universe is just a gigantic oscillator.

Entropy; Is Second Law of Thermodynamics Reversible?

The entropy of a system, σ, is equal to the natural logarithm of the volume of phase space accessible to the system. I think this derivation is from a textbook, but I don't remember which one.

$$(54) \qquad\qquad \sigma = \ln\Delta\Gamma.$$

$\Delta\Gamma$ is a volume of N particles in phase space, having units of mass times length times velocity, $(mxv)^{3N}$, that is, $action^{3N}$. Therefore,

$$(55) \qquad\qquad n = \Delta\Gamma/h^{3N},$$

where h is Planck's constant. So,

$$(56) \qquad\qquad \ln n = \ln\Delta\Gamma - 3N\ln h.$$

Since $3N\ln h$ is a fixed number, set it equal to σ_0, and, $\ln n = \sigma$, then

$$(57) \qquad\qquad \sigma + \sigma_0 = \ln\Delta\Gamma,$$

with the dimension of ln(action). Thus, we can say,

$$(58) \qquad\qquad \partial\sigma = \partial\ln\Delta\Gamma.$$

$\Delta\Gamma/h^{3N}$ is interpreted as a measure of randomness, so, σ, a number, is also a measure of randomness. Given a Maxwell-Boltzmann distribution of molecular velocities, note that the distribution starts at zero velocity and increases to a peak value then decreases to the maximum velocity, the speed of sound at the given density, pressure , and temperature of the gas. The peak location is related to the temperature. If we fix the temperature and volume of a perfect gas, $pv = nkT$, then as the density is increased, the pressure must also increase. Since the speeds are fixed between zero and the speed of sound, and the peak speed (the most likely value) is fixed by the temperature, the only thing that can change in the distribution as density and pressure increase, is the variance. For the Maxwell-Boltzmann distribution, variance is not symmetrical about the most probable value. One can look at the sum of the left and right deviations however, and see that as the density increases the peak value becomes more probable and the sum of the variances decreases. Hence, the total variance decreases as the pressure increases.

Now let the temperature rise as well. The peak moves to the right as well as increasing. The total variance approaches zero as the peak velocity approaches the maximum velocity. Almost all the molecules have the most probable speed which is now essentially the average speed and the root mean square speed as well. The distribution is approaching a normal one, with a symmetrical standard deviation. At a critical temperature and pressure when nearly all molecules are at their maximum allowed speeds, their freedom of motion has been reduced almost to zero. This is an unstable condition in which entropy is a minimum and potential energy has reached a maximum. Kinetic energy has now been stored like a compressed spring. If the particles are xions, the energy is stored in the strain of the particle represented by its mass.

The driving force for increasing xion density, pressure and temperature is provided by gravity. The gravitational constant changes, reaching a minimum when the system is most compressed. This allows the system to release it pent up energy in the form of radiation and explosive expansion. It is like a superheated fluid that suddenly explodes violently into an expanding gas.

As the system cools, the xions will condense, first to a liquid, then a solid. Defects now form, generating the familiar subatomic particles of our physical world. The gravitational constant is increasing even as the universe continues to expand and cool. Atoms form again and condense into stars. The universe is beginning to look more familiar.

At the opposite end of the cycle, the temperature, pressure, and density are very low, but the gravitational constant is approaching its maximum value. When it reaches that point the entropy of the system is at its maximum also. The spring is extended to its maximum and kinetic energy has virtually vanished. The system is again in a maximum potential energy state. The force between particles has become very large because of the effect of the gravitational constant. Recall that G is proportional to velocity squared and Planck's length squared divided by mass in equation (34) page 34. When the length is large and the mass small, as is the case when the xion lattice is cold, G will be large. Gravitational energy now starts transferring potential to kinetic energy as the system starts contracting. Pressure, temperature, and density of the xion material and its defects are increasing again and entropy is decreasing

Decreasing entropy in the universe means the Second Law of Thermodynamics has been reversed. If one can increase the kinetic energy in space, entropy will be lowered. That is because the variance in the phase space distribution of particles filling space has been decreased. Xions possess rest mass, as the author has suggested earlier. We know that the universal expansion will stop, if there is enough mass in the universe. So far the total calculated amount has fallen short of the required value, but it may not include all the mass of the xions. When that mass is added, the required number would probably be attained, then the expansion would stop and compression would begin. This is the point at which entropy would start to decrease; reversing the Second Law of Thermodynamics.

The universe resembles a Black Hole that contains the entire mass of that universe. To reiterate, the compression will continue until $\sigma + \sigma_0 \approx 0$. That means the randomness has been essentially removed from the system. Every xion will have the essentially same energy. The system is now unstable. There is no way to release the intense pressure in the xion material, so it must disintegrate. The entire system will release the stored energy in a huge inflation of the gas, since G is essentially zero. As it inflates, it will cool, allowing atoms to form again. Another cycle has begun. The constants of the universe are changing quickly in this phase. Some people have made a big case over the fact that certain constants are just what are needed for life to exist in the universe. I would suggest that the constants have changed and life was not possible until they reached certain values. Certainly the speed of light has changed as has the gravitational constant and the fine structure constant. True constants are like π, the square root of two and the base of natural logarithms. They may be rational or irrational, in either case they are connected to the ideal world of ideas, not the real one we live in.

Will the system repeat itself exactly? There is no reason why it should. It may repeat approximately. Galaxies may again form with their stars, planets, moons and comets. Life may begin again, but there is no guarantee that this solar system will be re-created exactly. There is no guarantee of mammalian intelligent life or the life of any individual being repeated. Intelligence will certainly evolve again, but the most intelligent being may not resemble a human being next time around.

There may be cycles however, in which no life is possible. In an infinite number of cycles, every conceivable universe is possible. Instead of the Multiverse suggested by some, in which innumerable universes coexist, one can imagine a system in which every universe has its place in the random progression of universes. Another alternative is for repeating, virtually identical universes, in which intelligent life creates an omniscient intelligence. In this scenario, the universe is the life cycle of a god(s), or Philosopher Kings. Their lives begin with the "big bang" and end with the "big crunch". How quickly speculation leads one to stray from science to philosophy and religion.

When entropy is a minimum, order is a maximum. When entropy is a maximum, the system is most random. Order and randomness are two aspects of the same thing, so to speak. Potential energy can be associated with order, and kinetic energy with randomness. Our Universe is just a complex oscillator or pendulum that forever swings between order and chaos. Order generates chaos, chaos generates order. We associate order with life, chaos with death, but the system does not seem to care, it is just doing its thing. Perhaps Born's headstone equation is a representation for the entire universe, just a great big oscillator, possibly only one of a countless number. It could be that the very small system and the very large only differ in scale, but these musings belong in the realm of science fiction.

Summary and Conclusion

This is a good place to stop, before the speculation gets out of hand. Many will say, it already has, no doubt. They will consider the whole idea of a new ether pure speculation. For them I offer this challenge; solve the twin paradox without appeal to a fixed reference frame. And let's not forget that vacuum pervaded by an electric field is in a state of stress and contains energy proportional to the square of the electric field strength. Shouldn't we inquire about what is being stressed? Empty space doesn't provide an answer.

Quintessence, the fifth state of matter has an interesting and continuing history. This extended article has offered the idea that xions are the constituents of that universal material. There is no doubt that one can derive quantum mechanics from a particle field model. The question is, what are those particles? Because of the ubiquity of photons in nature, one might consider them except for the lack of mass and low density. Whatever it is, it should be both particle and wave according to quantum theory. I don't accept this claim. The confusion, I believe, arises from mixing the particle, the xion, with its random oscillation in an ensemble of identical xion particles. It is as if the water wave and the water molecule were thought to be two aspects of the same thing. We know that the water molecule must move in a periodic fashion in order to generate a wave, but the wave is most certainly not part of the molecule. This confusion probably stems from the de Broglie suggestion of matter waves and their demonstration by Davisson and Germer. When they showed electron interference patterns resulting from the passage of high energy electrons through thin silver films, everyone was convinced that waves and particles were two aspects of the same thing. Born's interpretation of these waves as probability functions did not change that perception. University professors will tell you that electrons do indeed wave through the lattice. The idea that a probability function is a purely mathematical construction is lost here. They believe the particle has become a wave. It is not so. There is no wave-particle duality.

In the case of electrons, the pattern depends on the fact that the particles are charged and carry their fields with them. Even if the electrons are sent through the crystal lattice one at a time, the location of a single charged particle affects the next one's position, because it is influenced by the predecessor's field. The locations of all previous electrons affect the position of the next one. There is no magic here and no wave-particle duality either.

Xions are certainly particles, collectively they generate and transmit waves. Think about a toy composed of ten identical metal spheres attached to a rail by ten identical strings. The first sphere is pulled up to some height and released like a pendulum arm. All the spheres, which just touch each other, remain at rest except for the last one which completes the pendulum action. The energy of the falling ball is transmitted through the row of spheres as a wave, but you don't detect it, you detect the ball at the end which magically swings up. Would you say the ball you saw swing at the end was a wave? No. If however you arranged a rigidly fixed diaphragm to contact the last ball, it would oscillate and generate a wave you could detect. Thus, if an experiment was set up to look for a wave that is what it would find. If it was set up to detect a particle, that is what it would detect. No mystery and no wave- particle duality.

How is it possible for anyone to read the argument presented for the origin of quantum mechanics in the kinetic theory of a xion material and not be convinced? The evidence for the FitzGerald-Lorentz interpretation of their contraction hypothesis is just as convincing. Given the twin paradox, there is no other solution.

Proposing a xion material is certainly speculative. No more so than the, "branes" in eleven dimensions, of string theory, however. And what about that criterion – a theory is acceptable as long as it works? The dynamic crystal has several attractive features that tentatively seem to "work". Planck's length and time find a happy home in the lattice. We don't have to think of space-time dissolving into messy foam anymore. What could fit the idea of a field better than a stressed lattice able to support transverse elastic waves? Years ago in a Utica College physics class, Professor Peter Fong asked his students this rhetorical question, "Why are all electrons alike"? The xion lattice may finally provide the answer, they are an emergent property of the xion crystal.

Charges make sense as centers of mechanical stress. Rest mass and gravity have such natural explanations in this model, that it is difficult to believe there is not something substantial here. In addition, it puts you in great company, remember Sir Isaac Newton.

The model has far reaching power to provide sensible answers. It feels right. Specifics are something else. The devil is in the detail.

A quark, as a quantized stress state, is certainly speculation. Maybe it has merit, maybe not, but as Penrose said, we need some new ideas, perhaps even a new paradigm. Motion of an object as covariant diffusion through a particle based medium is a new way to look at some old problems. It certainly provides a unifying perspective. Many existing, unexplained facts have obvious solutions in the context of a xion material, for example, the mysterious force slowing the pioneer space craft, and quintessence. The unexpected observation that the universal expansion rate is increasing has a simple explanation when seen from this point of view.

The interpretation of tunneling provided by this model is especially notable. The traditional explanation has almost magical aspects. Radioactive decay likewise is easily understood in the context of a universal material. The extended lifetime of muons, often cited as proof of special relativity, is so readily explained in terms of an accelerated clock, as to be almost too easy, too logical, too compelling.

Most speculative is the last section, which ponders the problem of how this wonderful organism we call the universe got started. The "Big Bang" theory was a welcome development for the monotheistic religions, finally some good news after centuries of battering by this upstart, natural science. The Big Bang seems to require a creator, an intelligent designer. But that just puts off the problem because the curious and skeptical immediately inquire - where did He, She, or It come from? The natural solution is an oscillating universe. Then we could say, this is just the way nature is and always has been. But that still leaves the ultimate question – Why?

For nearly thirty years, the author has had an attention getting idea – quantum mechanics and relativity can both be derived by considering the universe to be filled with a material and applying our existing knowledge of material behavior. Attempts to present these ideas to the scientific community have been thwarted however, because publishers are so wary of publishing crank articles. They automatically turn down any proposed article that deals with the ether, unless of course it is written by a well known, published author. It is just too difficult to actually read every manuscript and make the judgment on technical merit. Sending an unsolicited article to a well known scientist is also usually a dead end. These people get more than their share of crank correspondence. So the best way to get published is the traditional one, if unknown, be associated with a university or research institution and submit a traditional article which offers no argument with accepted ideas. Until someone with an established reputation publishes the idea or is willing to sponsor yours, you are pretty much forced to join the fringe science community or publish your own book and hope someone discovers it after your ideas have become acceptable. The author has chosen the second approach and will trust his reputation to chance. It would be wonderful to be recognized, but science will move forward regardless. Truth will out.

References:

1. R. B. Laughlin, A Different Universe, Basic Books, NY, 2005

2. N.T.Greenspan, The End of a Certain World, Basic Books (2005)

3. A.A. Penzias, & R.W.Wilson, "A Measurement of Excess Antenna Temperature at 4080Mc/s", Astrophysical Journal, 142:419-421, July '65

4. R.Penrose, The Road to Reality, Alfred Knopf, NY, 2004

5. B. Rosenblum and F. Kuttner, Oxford University Press, 2006

6. C.F. Stevens, The Six Core Theories of Modern Physics, A Bradford Book, MIT Press, Cambridge, MA (1965) p152

7.. McClintock and Argon, Mechanical Behavior of Materials, Addison-Wesley Publishing Co. 1966,p110

8. R.W.Cahn, Physical Metallury, North-Hooland Publishing Co. Amsterdam, 1965, p630

9. C.W.Misner,TK.S.horne, & J.A.Wheeler, "Gravitation" W.H.Freeman & Co., San Francisco, CA, 1973, p 12

10. A. Einstein, Investigations on the Theory of Brownian Movement, Dover Publications, 1956

11. A. Aczel, Descartes' Secret Notebook, Broadway Books, NY 2005

12. S.K. Lamoreaux & J.R. Torgerson, "Neutron Moderation in the Oklo Natural Reactor and Time Variation of Alpha" Physical Review D69(2004): 121701-6

Note: The above reference is from pages 139-140 of the book, The Singularity is Near, by Ray Kurzweil, Viking, 2005

Appendix I

History of the Ether

The concept of ether was contained in Sanskrit [1, 2] writings. The Upanishads mention Akasha as the principle from which everything else results. This idea was apparently widespread in the ancient world. The hierarchy of energies in Hindu thought, are exactly the same and in the same order as the Greek – ether, fire, air, water, and earth. The Greek word for ether referred to an energy more fundamental, and therefore more pervasive than fire. The word was used by Homer and Hesiod to describe the stuff of heaven.

In more modern times, Rene' Descartes[3] used the concept of ether to explain how celestial bodies could be supported in an otherwise empty space and how action could be transmitted between bodies not in contact. Descartes could not accept the idea of "action at a distance". It was illogical, foreign to the course of scientific thinking inherited from the Greeks. Thus, although modern science traces the ether idea to Descartes, it actually pervaded western thought for thousands of years.

Since the time of Descartes, ether has been used by thinkers to explain various phenomena. Hooke [4] employed it to explain color as ether oscillations. Huygens [5, 6] and Torricelli [6] expanded on a Descartes idea and generated the luminiferous ether to explain various optical phenomena via mechanical analogies.

Although Newton had originally used the idea of action at a distance in his theory of gravity, he was also responsible for formally introducing ether into classical physics[7], stating that there is an aethereal medium, much like air but more elastic, capable of supporting light induced vibrations. In later years, Newton reconsidered action at a distance and discarded it, proposing that gravity was the result of density gradients in the ether. In the sometimes heated discussions between proponents of contiguous action, Cartesians, and those of action at a distance, Newtonians, the parties seemed only to have been aware of Newton's earlier position.

Many proposed ethers existed at one time. Hands down, the most famous was the luminiferous variety. Euler asserted that all electrical phenomena were manifestations of the luminiferous ether. Maxwell demonstrated the validity of this idea, but the success of

Newton's gravitational law and the unsatisfactory nature of Descartes' vortices in a continuous medium, led to the temporary demise of the ether in 1777.

The concept was revived when the wave theory of light gained sway in the early 19th century. An elastic material was required because of the transverse nature of light waves. Actually, it seemed that only a rigid solid could support such a wave, but that was impossible. The search for an explanation spun off developments in fluid mechanics and the elasticity of solids. Faraday suggested that the ether must support magnetic as well as electric forces. Maxwell's equations of electrodynamics are mathematical statements of Faraday's many original ideas in electricity and magnetism, including the notion of a field. Maxwell noticed the wave nature of the solutions to those equations, implying that the luminiferous ether and the electromagnetic ether were one in the same material.

Michelson and Morely[8] tried to lay the argument to rest with a measurement of the ether wind, if it existed. Their null result only caused more argument and speculation. Both FitzGerald, an Irish physicist, and Lorentz proposed that the moving arm contracted in the direction of motion. Note that the Doppler Effect could not detect a wind because the light source and the receiver are fixed in space relative to each other. Because of difficulties with the exact nature of the ether as discussed by Lorentz, FitzGerald modified his solution, suggesting that material bodies in motion contract relative to the ether. Thus, the famous Lorentz-FitzGerald contraction hypothesis, which became part of the Lorentz transformation equations proved the existence of the ether. But the theory was an ad hoc one which was viewed with suspicion.

Einstein's theory of relativity asserted that only relative motion of objects could be observed, not the motion relative to the ether. This idea was interpreted by others to mean that special relativity disproved the ether model. This is not what Einstein seemed to believe however. Einstein said [9], "Careful reflection teaches us that special relativity does not compel us to deny ether". The isotropic nature of the universe, as required by the Einstein model, has been called into question [10] by measurements of the microwave background remaining from the "big bang".

There is a compelling reason to retain the ether idea. The properties of space, e.g. permittivity, permeability, impedance, a maximum speed for information transmission, etc., are properties of something. It makes no sense to to talk about the physical properties of nothing. Newton's absolute space could well have been called ether. Einstein used the words, "empty space". Some refer to space-time. They could just as well use the word ether. Einstein pointed out that Mach's Principle, mass resides not in an individual body, but is the product of the entire universe, requires an ether if action at a distance is unacceptable. Such ether affects individual particle mass, but particle mass also affects the condition of the ether. Obviously, Einstein was influenced by Mach's idea, since it appears in General Relativity. The ether defined by relativity requirements, is a medium devoid of mechanical properties itself, but one which determines mechanical events, including electricity and magnetism. Einstein said he didn't need ether for special relativity, but in general relativity he succeeds in unifying all the ethers of scientific history. General Relativity could be called the Unified Ether Theory. Einstein never denied ether's existence really; rather he seemed to view it as a substance with some definite properties. Similarly, Pauli [11] described the ether as the totality of properties associated with matter free space.

Certainly, the elastic continuum envisioned in an earlier period is gone, but the concept of an all pervading medium has never been disproved. In fact, quantum mechanics requires vacuum be something and we now know photons permeate all space. The author demonstrated in a 3Apr79 paper [12] that if one considers the ether a medium of sub electron mass particles in rapid random motion, relativity and quantum mechanics are readily unified. Unfortunately, he was never successful in getting the paper published. A reviewer for Speculations in Science and Technology, in his rejection, noted that if space were filled with particles, they would have to be photons. Curiously, another paper the author submitted to the same publication, on a whim, was accepted. It should not have been, since the author, at the time, clearly did not understand the dynamo that is responsible for the earth's magnetic field. Obviously, acceptance or rejection was not based upon technical content, but rather the fact that one article suggested the ether was real. The "ether" paper was rejected, the technically faulty one was published.

Refernces:

1. Ramachataka, Y., The Philosophies and Religion of India, Yogi Publication Soc.   Chicago, Ill. P359 (1936)
2. Taimni, I.K., The duality of time and space, The Theosophist, 94, 256-272 (1973)
3. Descartes, R., The Geometry of Rene' Descartes, Dover Pubs xiii+244(1954)
4. Hooke, R., Micrographis- or Some Physiological Descriptions on Minute Bodies   Made by Magnifying Glasses, Martyn and Allentry, London, p246(1665)

5. Huygens, C., Treatise on Light, Dover Pubs xii+129(1962)
6. Whittaker, E.T., A History of the Theories of Aether and Electricity, The Classical Theories, Thomas Nelson and Sons Ltd., London(1958)
7. Newton, I., Phil Trans Roy Soc., London, tract no. 80, 3075-3087 and 5084-5103(1672)
8 .Michelson, A.A. and E.W. Morley, On the relative motion of the earth and the luminiferous ether, Am. J. Sci., 39, 333(1887)
9. Einstein, A., Aether and Relativitatstheorie, Rede gehalten am 5. Ma. 1920 (University of Lieden)Bln, Julius Springer (1920)
10.Rowan-Robinson, M., Aether drift detected at last. Nature, 270 9(1977)
11.Pauli,W., Theory of Relativity, Pergamon Press, London(1958)
12.Lane, C.H., Spacetime, A Sub electron Particle Field, 3Apr79 Unpublished

Acknowledgement: This article was written in May 1980 using material from an article in Speculations in Science and Technology entitled, Ether, Its Origin and History in Western Thought, as well as other sources.

www.ingramcontent.com/pod-product-compliance
Lightning Source LLC
Chambersburg PA
CBHW021859170526
45157CB00005B/1882